5th edition

geog.3

geography for key stage 3

< rosemarie gallagher >
< richard parish >

OXFORD
UNIVERSITY PRESS

OXFORD
UNIVERSITY PRESS

Great Clarendon Street, Oxford, OX2 6DP, United Kingdom

Oxford University Press is a department of the University of Oxford. It furthers the University's objective of excellence in research, scholarship, and education by publishing worldwide. Oxford is a registered trade mark of Oxford University Press in the UK and in certain other countries

© RoseMarie Gallagher, Richard Parish 2021

The moral rights of the authors have been asserted

Database right of Oxford University Press (maker) 2021

First published in 2001
Second edition 2005
Third edition 2009
Fourth edition 2015
Fifth edition 2021

All rights reserved. No part of this publication may be reproduced, stored in a retrieval system, or transmitted, in any form or by any means, without the prior permission in writing of Oxford University Press, or as expressly permitted by law, by licence or under terms agreed with the appropriate reprographics rights organization. Enquiries concerning reproduction outside the scope of the above should be sent to the Rights Department, Oxford University Press, at the address above.

You must not circulate this work in any other form and you must impose this same condition on any acquirer

British Library Cataloguing in Publication Data
Data available

ISBN 978-0-19-848991-7

10 9 8 7 6 5 4 3 2 1

MIX
Paper from responsible sources
FSC
www.fsc.org
FSC® C007785

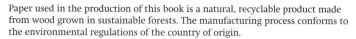

Paper used in the production of this book is a natural, recyclable product made from wood grown in sustainable forests. The manufacturing process conforms to the environmental regulations of the country of origin.

Printed in Great Britain by Bell and Bain Ltd., Glasgow.

Acknowledgements
The publisher and authors would like to thank the following for permission to use photographs and other copyright material:
Cover: OUP/Shutterstock. **Photos: p5:** Sven Broeckx/500Px Plus/Getty Images; **p6(t):** TMI / Alamy Stock Photo; **p6(bl):** NATURAL HISTORY MUSEUM, LONDON/Science Photo Library; **p6(bm):** Tyler Boyes/Shutterstock; **p6(br):** Art Directors & TRIP / Alamy Stock Photo; **p7(tl):** Andrew Lam/Shutterstock; **p7(tr):** Nigel Greenstreet / Alamy Stock Photo; **p7(ml):** travellight/Shutterstock; **p7(mr):** WAYHOME studio/Shutterstock; **p7(bl):** FILBERT RWEYEMAMU/AFP/Getty Images; **p7(br):** Patrick AVENTURIER/Gamma-Rapho/Getty Images; **p8(t):** dragonjian/Shutterstock; **p8(b):** beboy/Shutterstock; **p9(tl):** Sinclair Stammers/Science Photo Library; **p9(tr):** John Reader/Science Photo Library; **p9(bl):** Dr Keith Wheeler/Science Photo Library; **p9(br):** Adrian Sherratt / Alamy Stock Photo; **p10(t):** nagelestock.com / Alamy Stock Photo; **p10(ml):** T-Design/Shutterstock; **p10(m):** Global Warming Images/Shutterstock; **p10(mr):** Gary Crabbe / Enlightened Images / Alamy Stock Photo; **p10(bl):** Steven Milne / Alamy Stock Photo; **p10(bm):** Gregory Dimijian / Science Photo Library; **p10(br):** eye-for-photos/Shutterstock; **p11(tl):** Pi-Lens/Shutterstock; **p11(tm):** serato/Shutterstock; **p11(tr):** katarinag/Shutterstock; **p11(b):** Oliver Smart / Alamy Stock Photo; **p12:** Filip Fuxa/Shutterstock; **p13(t):** ako photography/Shutterstock; **p13(b):** PA Images / Alamy Stock Photo; **p14:** mileswork/Shutterstock; **p15:** PRAWNS / Alamy Stock Photo; **p16(b):** Richard Bradford/Shutterstock; **p17(t):** John Bentley / Alamy Stock Photo; **p17(b):** East Anglia Images / Alamy Stock Photo; **p18(t):** Derek Stone / Alamy Stock Photo; **p18(bl):** Paul Miguel / Alamy Stock Photo; **p18(br):** RichardBaker/Alamy Stock Photo; **p19:** John W Banagan/Getty Images; **p20:** blickwinkel / Alamy Stock Photo; **p21(tl):** Andy Aleks/Shutterstock; **p21(tm):** sirtravelalot/Shutterstock; **p21(tr):** Lukas Gojda/Shutterstock; **p21(mr):** djgis/Shutterstock; **p21(m):** Jacek Chabraszewski/Shutterstock; **p21(ml):** Prostock-studio/Shutterstock; **p21(b):** Jan Fritz/Alamy Stock Photo; **p22(t):** Fotema/Shutterstock; **p22(b):** Aneta_Gu/Shutterstock; **p23:** agefotostock/Alamy Stock Photo; **p25(t):** Ian_Stewart/Shutterstock; **p25(b):** Tom Hanslien Photography / Alamy Stock Photo; **p26(t):** Photoagriculture/Shutterstock; **p26(b):** AdeleD/Shutterstock; **p28(l):** Jim Gipe / Agstockusa / Science Photo Library; **p28(r):** PhotoQuest/Getty Images; **p29:** inga spence / Alamy Stock Photo; **p30(l):** Laurent Weyl/Argos/Panos Pictures; **p30(tr):** courtneyk/iStock/Getty Images; **p30(m):** David J. Green / Alamy Stock Photo; **p30(bl):** Chandan kumar Bisai/Shutterstock; **p30(br):** Alchemist from India/Shutterstock; **p31(tl):** Mick Harper/Shutterstock; **p31(tr):** REUTERS / Alamy Stock Photo; **p31(b):** ZUMA Press, Inc. / Alamy Stock Photo; **p32(t):** Peo Quick / Alamy Stock Photo; **p32(b):** Noam Armonn/Shutterstock; **p33(t):** Eat Just, Inc.; **p33(b):** In Pictures Ltd./Corbis/Getty Images; **p34(t):** Africa Media Online / Alamy Stock Photo; **p34(b):** REUTERS / Alamy Stock Photo; **p35(t):** adike/Shutterstock; **p35(t):** 35mmf2/Shutterstock; **p35(b):** Sue Cunningham Photographic / Alamy Stock Photo; **p36(t):** Joerg Boethling / Alamy Stock Photo; **p36(m):** PICS Global; **p36(bl):** International Crops Research Institute for the Semi-Arid Tropics (ICRISAT); **p36(br):** Photononstop/Superstock; **p37(t):** Anton_Ivanov/Shutterstock; **p37(b):** ESA / ATG Medialab; **p38(tl):** hans engbers/Shutterstock; **p38(tm):** TK Kurikawa/Shutterstock; **p38(tr):** Koldunov/Shutterstock; **p38(b):** thi/Shutterstock; **p39:** dugdax/Shutterstock; **p40(tl):** nostal6ie/Shutterstock; **p40(tm):** Travel Stock/Shutterstock; **p40(tr):** Steve Meese/Shutterstock; **p40(bm):** Atlantis Resources Ltd.; **p40(br):** Tim Scrivener / Alamy Stock Photo; **p42(tl):** Justin Kase zsixz / Alamy Stock Photo; **p42(tr):** CHAIYA/Shutterstock; **p42(b):** © 2019 The World Bank, Source: Global Solar Atlas 2.0, Solar resource data: Solargis. **p43(tl):** robertharding/Alamy Stock Photo; **p43(tr):** SIA KAMBOU/AFP/Getty Images; **p43(b):** Courtesy of Gabriele Diamanti; **p44(t):** geogphotos/Alamy Stock Photo; **p44(ml):** blickwinkel / Alamy Stock Photo; **p44(m):** dpa picture alliance / Alamy Stock Photo; **p44(mr):** Minden Pictures / Alamy Stock Photo; **p44(b):** Erni/Shutterstock; **p45(tl):** David Cole / Alamy Stock Photo; **p45(tr):** 717Images by Paul Wood/Moment Open/Getty Images; **p45(b):** 1989studio/Shutterstock; **p46(t):** Protasov AN/Shutterstock; **p46(b):** JaxxLawson/Shutterstock; **p47:** Mike Kemp/In Pictures/Getty Images; **p48(t):** Geraint Lewis/Alamy Stock Photo; **p48(ml):** Haraldmuc/Shutterstock; **p48(mr):** Monty Rakusen/Cultura/Getty Images; **p48(bl):** Wavebreakmedia/Shutterstock; **p48(br):** Janine Wiedel Photolibrary/Alamy Stock Photo ; **p49(t):** Jim Richardson/Getty Images; **p49(b):** Andersen Ross/Photodisc/Getty Images; **p52(t):** Mark Lees/Alamy Stock Photo; **p52(m):** Edward Moss/Alamy Stock Photo; **p52(b):** Suzanne Plunkett/Bloomberg/Getty Images; **p53(t):** Commission Air/Alamy Stock Photo; **p53(b):** Image Source/Alamy Stock Photo; **p54(tl):** Daniel J. Rao/Shutterstock; **p54(tr):** TRMK/Shutterstock; **p54(bl):** Andrey_Popov/Shutterstock; **p54(br):** unoL/Shutterstock; **p55:** Ian Fairbrother/Stockimo/Alamy Stock Photo; **p56(t):** clubfoto/iStock/Getty Images; **p56(t):** rtguest/Shutterstock; **p56(ml):** Alex Kolokythas Photography/Shutterstock; **p56(m):** Song_about_summer/Shutterstock; **p56(tr):** Bokic Bojan/Shutterstock; **p56(b):** oasistrek/iStock/Getty Images; **p57:** Iain Masterton/Alamy Stock Photo; **p58(t):** SpeedKingz/Shutterstock; **p58(m):** Stringer Shanghai/Reuters; **p58(b):** Wang Lei/Imagine China/Shutterstock; **p59:** Robert Way/Shutterstock; **p60(tl):** Lordprice Collection/Alamy Stock Photo; **p60(tm):** incamerastock/Alamy Stock Photo; **p60(tr):** MagicBones/Shutterstock; **p60(bl):** A G Baxter/Shutterstock; **p60(bm):** Croisy/Shutterstock; **p60(br):** InPerspective/Shutterstock; **p61(tl):** Dragon Images/Shutterstock; **p61(tm):** luciano de polo stokkete/Alamy Stock Photo; **p61(tr):** Richardjohnson/Shutterstock; **p61(b):** SolStock/Getty Images; **p62(tl):** Matt Cardy/Getty Images News/Getty Images; **p62(tm):** Metamorworks/iStock/Getty Images; **p62(tr):** Starcevic/iStock/Getty Images; **p62(b):** Sharomka/Shutterstock; **p63(t):** 1000 Words/Shutterstock; **p63(b):** Joe Ferrer/Shutterstock; **p64(t):** Georgeclerk/iStock/Getty Images; **p64(b):** masy100/Shutterstock; **p65:** Katoosha/Shutterstock; **p66(tl):** Dennis Diatel/Shutterstock; **p66(tr):** Joerg Boethling / Alamy Stock Photo; **p66(ml):** PhilMSparrow/iStock/Getty Images; **p66(mr):** NellyUlusoy/Shutterstock; **p66(b):** CherylRamalho/Shutterstock; **p69:** anasalhajj/Shutterstock; **p68(l):** Thomas Imo/Photothek/Getty Images; **p68(r):** Tetyana Dotsenko/Shutterstock; **p71:** Jodi Baglien Sparkes/Shutterstock; **p73(t):** Sam Pollitt / Alamy Stock Photo; **p73(m):** Dietmar Temps / Alamy Stock Photo; **p73(bl):** JULIAN LOTT/Shutterstock; **p73(br):** erichon/Shutterstock; **p77(l):** REUTERS / Alamy Stock Photo; **p77(r):** imageBROKER / Alamy Stock Photo; **p78(t):** Zhao jiankang/Shutterstock; **p78(b):** Tero Vesalainen/iStock/Getty Images; **p79(t):** SHAMSUL HAIDER BADSHA/AFP/Getty Images; **p79(b):** Sk Hasan Ali/Shutterstock; **p80(t):** ifong/Shutterstock; **p80(b):** Nelson Antoine/Shutterstock; **p81(t):** Rafiq Maqbool/AP/Shutterstock; **p81(b):** © UNICEF/UN0421460/Kokoroko; **p82(t):** Kypros/Moment Unreleased/Getty Images; **p82(bl):** Marco Di Lauro/Getty Images; **p82(br):** PjrNews / Alamy Stock Photo; **p83(t):** Chess Ocampo/Shutterstock; **p83(m):** Jim West / Alamy Stock Photo; **p83(b):** Jim West / Alamy Stock Photo; **p84(tl):** manoj_kulkarni/Shutterstock; **p84(tm):** Sunday Alamba/AP/Shutterstock; **p84(tr):** Friedrich Stark / Alamy Stock Photo; **p84(bl):** Piotr Wawrzyniuk/Shutterstock; **p84(bm):** SEYLLOU/AFP/Getty Images; **p84(br):** Simon Rawles / Alamy Stock Photo; **p85(tl):** Scott Bairstow / Alamy Stock Photo; **p85(tm):** Tim Graham / Alamy Stock Photo; **p85(tr):** imageBROKER / Alamy Stock Photo; **p85(b):** Trevor Snapp/Bloomberg/Getty Images; **p86:** Jake Lyell/Alamy Stock Photo; **p87(t):** HuHu/Shutterstock; **p87(b):** Super Nova Images / Alamy Stock Photo; **p88(l):** dave stamboulis/Alamy Stock Photo; **p88(r):** SPUTNIK/Science Photo Library; **p89(l):** Badger Castle/Shutterstock; **p89(r):** Itsik Marom / Alamy Stock Photo; **p93(l):** NOAA/NCEI; **p93(r):** ©ReeveJolliffe; **p95:** Xinhua/Shutterstock; **p96(t):** Zhang Lei/Cns/EPA/Shutterstock; **p96(b):** Feng Li /Getty Images News/Getty Images; **p97:** Sipa/Shutterstock; **p98:** JOHN RUSSELL/AFP /Getty Images; **p99(t):** DigitalGlobe /Getty Images; **p99(b):** Archiv Mehrl/ullstein bild/Getty Images; **p100(l):** Juancat/Shutterstock; **p100(r):** Tom Pfeiffer / VolcanoDiscovery; **p101(t):** Jefta Images / Barcroft Media/Getty Images; **p101(b):** ATAR/AFP/Getty Images; **p102:** NASA/GSFC/METI/ERSDAC/JAROS and US/JapanASTER Science Team; **p103(t):** balounm/Shutterstock; **p103(b):** Antonio Capone/AGF/Universal Images Group/Getty Images; **p104(l):** Hamdan Yoshida/Shutterstock; **p104(r):** Marco Ossino/Shutterstock; **p105:** Jacek Nowak / Alamy Stock Photo; **p106(t):** Buyenlarge/Getty Images; **p106(b):** Cavan Images / Alamy Stock Photo; **p107:** Katsiaryna Pleshakova/Shutterstock; **p107(tl):** holgs/iStockphoto; **p107(tr):** ITAR-TASS News Agency / Alamy Stock Photo; **p107(ml):** ITAR-TASS News Agency / Alamy Stock Photo; **p107(m):** NikolayN/Shutterstock; **p107(m):** AridOcean/Shutterstock; **p107(mr):** Leonid Ikan/Shutterstock; **p107(bl):** FotograFFF/Shutterstock; **p107(br):** Andrey Rudakov/Bloomberg/Getty Images; **p109:** Budkov Denis/Shutterstock; **p110:** Igor V. Podkopaev/Shutterstock; **p111(t):** robertharding/Alamy Stock Photo; **p111(m):** Buddy Mays / Alamy Stock Photo; **p111(b):** Alexander Piragis / Alamy Stock Photo; **p112(t):** Nature Picture Library / Alamy Stock Photo; **p112(b):** Serg Zastavkin/Shutterstock; **p113(tl):** Maximillian cabinet/Shutterstock; **p113(tr):** PaulPaladin/Alamy Stock Photo; **p113(bl):** Huldiberdiev/Shutterstock; **p113(br):** Denis Ukolov/Shutterstock; **p114(l):** Andrey Rudakov/Bloomberg/Getty Images; **p114(r), p116(bl):** DEA / G. SIOEN/De Agostini/Getty Images; **p115:** Alexey anashkin/Shutterstock; **p116(t):** Yadid Levy/Alamy Stock Photo; **p116(bm):** Dmitro2009/Shutterstock; **p116(br):** Aleksander Karpenko/Shutterstock; **p117(tl):** ITAR-TASS News Agency/Alamy Stock Photo; **p117(tm):** Vitaliy Kaplin/Shutterstock; **p117(tr):** ITAR-TASS News Agency / Alamy Stock Photo; **p117(bl):** Tatyana Mi/Shutterstock; **p117(br):** Nadia Isakova/Alamy Stock Photo; **p118(t):** Dean Conger/Corbis/Getty Images; **p118(m):** Bryan and Cherry Alexander/Nature Picture Library; **p118(bl):** Tatiana Gasich/Shutterstock; **p118(bm):** Mikhail Cheremkin/Shutterstock; **p118(br):** FotoSoyuz/Hulton Archive/Getty Images; **p119(t):** Michael Robinson Chavez/The Washington Post/Getty Images; **p119(m):** MLADEN ANTONOV/AFP/Getty Images; **p119(b):** TASS/TASS/Getty Images; **p120:** Nature Picture Library / Alamy Stock Photo; **p121(t):** AP Photo/RTR Russian Channel/Shutterstock; **p121(b):** ITAR-TASS News Agency / Alamy Stock Photo; **p122(t):** Padi Prints / Troy TV Stock / Alamy Stock Photo; **p122(b):** Andrey Deryabin / Alamy Stock Photo; **p123(h):** Greens87/Shutterstock; **p123(h):** Macrovector/Shutterstock; **p123(h):** zokka/Shutterstock; **p123(tl):** Ischmidt/Shutterstock; **p123(tr):** imageBROKER / Alamy Stock Photo; **p123(ml):** Ayhan Altun / Alamy Stock Photo; **p123(m):** Universal Images Group North America LLC / Alamy Stock Photo; **p123(mr):** Saudi Desert Photos by TARIQ-M/Moment/Getty Images; **p123(bl):** Alexander Ishchenko/Shutterstock; **p123(br):** STR/AFP/Getty Images; **p124:** Justin Kase zsixz / Alamy Stock Photo; **p125(t):** Belikova Oksana/Shutterstock; **p125(b):** Rasto SK/Shutterstock; **p126:** Michal Knitl/Shutterstock; **p127(t):** Oleksandr Kalinichenko/Shutterstock; **p127(b):** Jesse Allen/NASA EQ-1 team/NASA Earth Observatory; **p128:** Alberto Loyo/Shutterstock; **p129(tl):** LakedemonPhoto/Shutterstock; **p129(tr):** Rafael Ben-Ari / Alamy Stock Photo; **p129(bl):** bahadir ay/Shutterstock; **p129(br):** Vlad61/Shutterstock; **p132(t):** Michele Burgess / Alamy Stock Photo; **p132(b):** NAPA/Shutterstock; **p133(t):** Joshua Stevens/NASA Earth Observatory; **p133(b):** Joerg Boethling / Alamy Stock Photo; **p134:** ehasdemir/Shutterstock; **p135:** LouieLea/Shutterstock; **p136:** VolodymyrT/Shutterstock.

Artwork by Q2A Media Services Pvt. Ltd, Mike Parsons (Barking Dog Art), Ian West, Mike Phillips, Szilvia Szakall (all Beehive Illustration), Simon Tegg (Simon Tegg Illustration), Our World in Data (CC BY 4.0), Mike Connor, Steve Evans and Tracey Learoyd (Oxford University Press).

The Ordnance Survey material on p 19 is reproduced with the permission of the Controller of Her Majesty's Stationery Office © Crown Copyright. The data on pp 39 (Global primary energy consumption by source, 2019: Our World in Data / Hannah Ritchie) and 71 (Human Development Index map for 2017) comes from Our World in Data and is reproduced under the Creative Commons BY-SA 3.0 AU License.

Ordnance Survey (OS) is the national mapping agency for Great Britain, and a world-leading geospatial data and technology organisation. As a reliable partner to government, business and citizens across Britain and the world, OS helps its customers in virtually all sectors improve quality of life.

Thanks to the website *Our World in Data*, a rich resource for data, maps, and inspiration, and to Stephanie Blott of FTI Consulting LLP, Helen Wilson-Prowse at Wave Hub Limited, and Kelly Downey at Shamin Abas.

The changes in this edition of geog.3 are the result of comments from many people. We would like to thank the teachers who came together in focus groups to discuss the course, as well as our reviewers who provided thoughtful and constructive criticism, and in particular Kate Stockings. A special thanks to Garaeth Davies.

Note that the content of any direct speech attributed to characters in this book is based on information from reliable sources.

Links to third party websites are provided by Oxford in good faith and for information only. Oxford disclaims any responsibility for the materials contained in any third party website referenced in this work.

Every effort has been made to contact copyright holders of material reproduced in this book. Any omissions will be rectified in subsequent printings if notice is given to the publisher.

Contents

Where have you got to?

1

From rock to soil

What exactly is rock? Find out here.

What is rock?

Rock is the hard material that forms Earth's surface, and most of Earth below the surface.

Earth's rock first formed over 4.5 billion years ago, when the dust and gases swirling around our Sun were pulled together by gravity. A ball of molten rock formed. It began to cool and harden. It is still cooling!

Always changing!

Rock looks permanent. But don't be fooled. It is changing all the time – usually very slowly. It gets broken into bits. Bits get stuck together to make new rock. It gets pulled deep into Earth. Some gets melted.

So what is rock made of?

Rock is a mixture of **minerals**. A mineral is a natural compound. It has a chemical name and formula, like the ones you meet in science – but we usually use its *geological* name.

Minerals usually exist in rock as **crystals**.

Look at these samples of rock. The different minerals have different colours.

Can I?

▲ *A granite rock face: a challenge to climbers.*

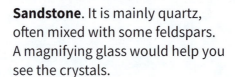

Granite. It is a mixture of minerals. The palest blobs are crystals of silicon dioxide or **quartz**. Most of the other minerals in it are **feldspars**.

Sandstone. It is mainly quartz, often mixed with some feldspars. A magnifying glass would help you see the crystals.

Limestone. It is mainly **calcite**, or calcium carbonate, often mixed with quartz and other minerals. **Chalk** is a form of limestone.

The most common minerals

There are around 5000 known minerals. But only about ten are common. These are called the **rock-building minerals**.

The most common are the feldspars, and quartz. Between them, they make up most of Earth's **crust** – the hard outer layer that you live on. There are several different feldspars. Like quartz, they contain silicon and oxygen, but also aluminium, and other elements such as sodium, calcium, and iron.

Did you know?

● The silicon for the chips in mobiles is made from sand that's almost pure quartz.

Why …

… does rock have different colours?

What do we use rock for?

Every year, we cut billions of tonnes of rock from Earth's crust. Why?

For building things

The building industry needs rock.

Some buildings are built from blocks cut from rock. **Concrete** is made of crushed rock and sand. **Cement** is made from limestone.

A layer of crushed rock or **aggregate** is used in road building.

To obtain metals

Metals come from rock. The rocks we get them from are called **metal ores**.

Some metals occur naturally as the metal, in rock. For example gold is found as lumps of gold.

But most metals occur within minerals. When the ores are dug up, the metals are extracted from the minerals using chemical reactions.

To obtain gemstones

Some very attractive minerals are found in very small quantities in rock. They are called **gemstones**.

Diamond, ruby, jade, and tanzanite, are examples.

Gemstones are used to make jewellery and other objects. Usually the rarer a gemstone is, the more expensive it will be.

▲ Concrete is the most-used building material in the world.

▲ A limestone quarry in Derbyshire. Some of the rock from here is used to make concrete.

▲ Aluminium is used in building planes. It is extracted from an ore called bauxite.

▲ Mobiles contain over 60 metals – all from rock! They include gold, silver, and tungsten.

▲ Tanzanite is found only in Tanzania. These chunks were sold for £2 million in 2020.

▲ When tanzanite is cut and polished, it forms deep blue stones like these.

Your turn

1 When did Earth's rock first form?

2 Rock is made of minerals. What are *minerals*?

3 a State the geological name for the mineral *silicon dioxide*.
   b Give two examples of rock that contain this mineral.

4 The rock *halck* is a form of limestone. Unjumble the word.

5 Give three ways in which the building industry uses rock.

6 a What are *metal ores*?
   b Name the ore from which we obtain aluminium.

7 Give five ways in which rock benefits *you*. Answer in any way you wish. (A list, a drawing, a spider map?)

8 Help! A fungus has arrived from outer space. It eats rock! Write an article for a news website about how this fungus is affecting the UK. At least 10 lines. Make it dramatic!

All rock belongs to one of three groups. Here you can find out about the groups, and how they differ.

Putting rock into groups

There are lots of different kinds of rock, with different mixtures of minerals. But they fall into just three groups of rock – **sedimentary**, **igneous**, and **metamorphic**. Each group has been formed in a different way. Let's look at each in turn.

Sedimentary rock

This forms when particles of minerals which were eroded from rock in one place get stuck together again in another place, to form new rock.

For example, a river carries a load of particles eroded from the rock it flows over. It carries them to the ocean and dumps them. They fall …

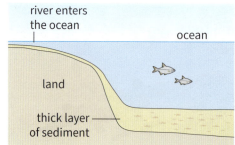

… to the ocean floor as sediment. Over years the layer of sediment builds up. It could be hundreds of metres thick, and very very heavy.

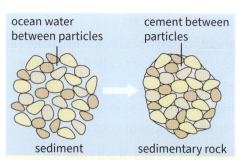

The weight causes the particles deep in the layer to get squeezed together. Substances dissolved in the water act as cement. The result: new rock!

Different sediments give different types of sedimentary rock. For example:

- a sediment of **mud** gives **mudstone**
- a sediment of **sand** gives **sandstone**
- a sediment of shells and coral piling up on the ocean floor gives **limestone**.

Igneous rock

This forms when rock melts below ground, then cools and hardens again.

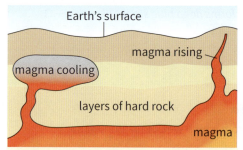

Deep below us, rock is very hot. Under certain conditions it melts, giving a liquid called **magma**. This rises and cools. When it cools slowly …

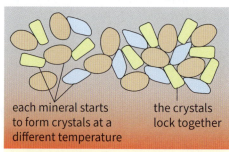

… underground, its minerals start to crystallise at different temperatures. The crystals grow, and lock together to give rock such as **granite**.

But some magma shoots out at volcanoes as **lava**. This cools quickly to form rock such as **basalt**, with small crystals.

Did you know?
- Limestone forms from shells and other hard remains of organisms, that pile up on the ocean floor.

Did you know?
- Chalk forms in the same way as limestone, but from very tiny organisms.

▲ 120 million years ago, in China, a young dinosaur got buried in sand. The sand formed sandstone. And someone found the fossil.

▲ 3.6 million years ago in Tanzania, Africa, a creature left footprints in volcanic ash. Later, the ash was compacted into a rock called **tuff**.

Metamorphic rock

This is rock which has been changed underground, *without* melting.

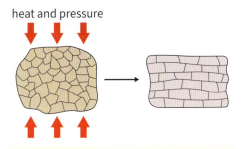

Deep underground, the heat and pressure can be enormous. Together, they can change any type of rock *without* melting it.

The structure of the minerals usually changes, so the rock looks different. For example this is the sedimentary rock **mudstone**. Underground, …

… heat and pressure cause it to metamorphose to **slate**, where the crystals are lined up in flat sheets. Slate is used for roof tiles.

Where are they found?

Where are those rock groups found? Look at this drawing of Earth's crust, the outer layer that you live on.

About 65 % of the crust is igneous rock. Only about 8 % is sedimentary. The rest is metamorphic.

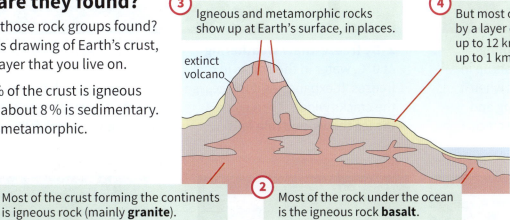

3 Igneous and metamorphic rocks show up at Earth's surface, in places.

4 But most of the surface is covered by a layer of sedimentary rock – up to 12 km thick on land, and up to 1 km thick under the ocean.

extinct volcano

Key
- sedimentary rock
- metamorphic rock
- igneous rock

1 Most of the crust forming the continents is igneous rock (mainly **granite**).

2 Most of the rock under the ocean is the igneous rock **basalt**.

Your turn

1 Most rivers carry **silt**, which is a mixture of small particles.
   a Draw a flow chart to show how and where silt becomes a sedimentary rock.
   b Suggest a name for the rock formed from silt.

2 a How are limestone and chalk formed? (*Did you know? …*)
   b Are limestone and chalk still forming? Give reasons.

3 Under high pressure, limestone will turn into marble.
   a To which group of rocks does marble belong?
   b How could marble be turned into an igneous rock?

4 a There's no sedimentary rock deep in Earth's crust. Why?
   b You won't find fossils in igneous rock. Suggest a reason.

5 Which rock group is the most common *at Earth's surface*?

Over time, all rock at Earth's surface is broken down by weathering. How does that work? Find out here.

What is weathering?

Look at the bare rock in the photo on the right. A few thousand years from now, it might have turned into stones and soil.

Why? Because of a set of processes called **weathering**.

In weathering, rock is broken down by the action of things in its environment: by heat and cold, rain, gases from the air, and even by plants and animals.

Two kinds of weathering go on together: **physical weathering** and **chemical weathering**.

▲ *Weathering in progress in Scotland. Very slowly!*

❶ Physical weathering

In physical weathering, the rock gets broken into bits – but the minerals in it *do not change*. Rock can be broken into bits in several ways.

1 By heating and cooling
Rock expands as it heats up in the sun, and contracts when it cools. Repeated heating and cooling can cause it to crack.

2 By freeze-thaw weathering
At 0 °C, water in the cracks in rock freezes. It expands as it freezes, so the cracks widen. Then when the ice thaws, and rain falls …

… the cracks fill with water again. It freezes again … and thaws again. Each time the cycle repeats, the cracks get even wider. Eventually the rock falls apart.

3 By a reduction in pressure
Rock deep underground is under pressure, due to the weight of the rock above it. But if the rock above is eroded away, it is no longer …

… under pressure. So it expands. This causes it to split, parallel to the surface. Over time, layers break off, like layers of an onion. This process is called **exfoliation**.

4 By living things
Roots work their way into cracks in rock, and widen them. Burrowing animals can also make cracks bigger.

② Chemical weathering

In chemical weathering, minerals in the rock undergo chemical change. This helps to weaken the rock, and break it up. Look at these examples.

gaps where the limestone has been dissolved

Rain is slightly acidic. That's because it dissolves carbon dioxide from the air. So it reacts with the calcium carbonate in limestone, and dissolves it.

The commonest minerals in rock are the **feldspars** (page 6). Water reacts with these over time, giving **clay**. This is made of very fine particles. It is soft, and slippery when wet.

Quartz is also very common – but it resists chemical weathering. So when rock containing quartz breaks down, the quartz crystals are set free as sand. They end up on our beaches!

The end result

Together, physical and chemical weathering break rock down into stones, and sand, and clay, and compounds which dissolve in water and are carried away.

When the sand and clay are mixed with rotting vegetation, the result is **soil**. You can find out more about soil in Unit 1.8.

How quickly?

How quickly does rock break down? It depends on the climate, and the rock.

- Daily temperature changes from hot to cold, or temperatures that vary around 0 °C, promote physical weathering.

- Chemical weathering speeds up as the temperature rises. And it usually needs rain. So it is fastest in a warm, damp climate.

- Some rock weathers more easily than others. For example, mudstone weathers much more easily than granite does.

So, depending on what and where it is, rock might form a few centimetres of soil in a few hundred years … or hardly any in a thousand.

lichen – a fungus combined with algae

▲ Some living things bring about both physical and chemical weathering. For example lichen grows into cracks in rock, and also makes acids which eat into rock.

Your turn

1 The set of processes which break down rock is called *weathering*. Suggest a reason why that name was chosen.

2 Page 10 shows four kinds of physical weathering. Choose *one*, and draw labelled diagrams to show how this process breaks rock into bits.

3 State *the key difference* between physical and chemical weathering.

4 Describe how clay is formed.

5 a One form of weathering shown on page 10 is often called *biological weathering*. Which one? Explain your choice.

  b Explain how rabbits can contribute to weathering.

  c Describe two ways in which lichen helps to break up rock.

6 Limestone weathers much more slowly in dry areas than in rainy areas. Explain why.

Sedimentary rock may change into metamorphic or igneous rock … then back to sedimentary rock. How? Find out here.

All change!

Rock moves from one rock group to another – and from ocean floor to mountain top. Amazing! It is summed up in the **rock cycle**, which goes on non-stop, 24/7, all over Earth. Look:

Did you know?
- The summit of Mount Everest is made of limestone ….
- … which formed on the ocean floor!

What if…
… the rock cycle stopped?

A

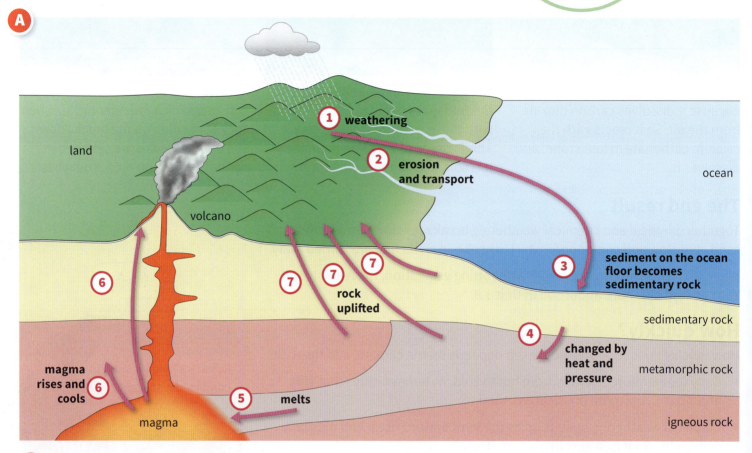

land
volcano
ocean
1 weathering
2 erosion and transport
3 sediment on the ocean floor becomes sedimentary rock
sedimentary rock
4 changed by heat and pressure
metamorphic rock
7 rock uplifted
6 magma rises and cools
5 melts
magma
igneous rock

① Rock at Earth's surface is not stable. It is broken down to stones and sand and clay by **weathering**.

② The river **erodes** (picks up) the stones and sand and clay, and **transports** them away. (Glaciers and the wind can also do this.)

③ As it enters the ocean the river **deposits** its load. A thick layer of sediment builds up on the ocean floor. It is compacted to form **sedimentary rock**.

④ Sedimentary rock gets forced down into Earth's crust. (You'll see why on the next page.) The heat and pressure change it to **metamorphic rock**.

⑤ Under certain conditions, the rock melts. The liquid rock is called **magma**.

⑥ The magma rises, and cools to form **igneous rock**. Some cools below Earth's surface. Some shoots out at volcanoes as **lava**, and cools at the surface.

⑦ Solid rock also gets raised or **uplifted**. The ocean floor, with its layer of sedimentary rock, is uplifted to form land. Igneous and metamorphic rock are uplifted too. When the uplifted rock reaches Earth's surface, it begins to weather. So the cycle starts all over again …

▲ David, sculpted from marble by Michelangelo. Marble is the metamorphic rock obtained from limestone. So David was once sea shells and coral!

What drives the rock cycle?

In the rock cycle, rock gets pushed down into Earth's crust (step 4 in **A**), *and* uplifted (step 7). How?

- First, you need to know that the hard outer part of Earth is cracked into huge slabs. We call these slabs **plates**.

- The plates are always on the move. They push into each other, and move away from each other, and slide past each other.

- Diagram **B** shows two plates pushing into each other. Plate X is heavier, so its edge sinks below the edge of plate Y. At the same time, the pressure causes uplift of rock on Y.

So it is **plate movements** like these which drive the rock cycle. You will learn more about plates in Chapter 5.

How does the rock cycle help us?

It helps us in two key ways.

1 Useful materials are formed.

For example, building materials like limestone (sedimentary rock), marble (metamorphic rock) and granite (igneous rock). And metal ores.

Some gemstones form in metamorphic rock, when minerals undergo change. Others form when superheated water dissolves minerals from magma. The solution cools in cracks in rock. As it cools, crystals of gemstone form.

Diamonds are different. They form about 150 km below us, in very hot igneous rock. They shoot out in magma erupting at volcanoes.

2 We gain access to these materials.

When rock is uplifted, or erupted from volcanoes, or when overlying material is eroded, we gain access to materials we want. For example, uplift can move limestone from the ocean floor onto land.

And also, very important: weathering leads to that essential material, **soil**.

B

▲ *Two plates pushing into each other. Some rock gets buried. Other rock gets uplifted.*

▲ *All thanks to the rock cycle!*

▲ *Thanks to the rock cycle! The British monarch wears this crown at the opening of Parliament.*

Did you know?
- The more slowly magma cools underground, the bigger the crystals will be.

Your turn

1 All rock on Earth's land surface will turn into *sedimentary rock* over time. Explain how this happens, with the help of **A**.

2 Put the six rocks below into matching pairs. Like this:
(sedimentary rock, then its matching metamorphic rock).
limestone   slate   quartzite
sandstone   mudstone   marble

3 The sediment that forms limestone is *not* carried into the ocean by rivers. So where does it come from?

4 What are *plates*, in geography?

5 Explain why plate movements can cause sedimentary rock to turn into metamorphic rock. **B** will help.

6 Limestone forms under water. So why is it found widely on land?

7 Write a summary of the rock cycle in no more than eight lines.

8 Look at the street scene above. Then explain the caption!

The land you are living on once lay at the Equator. How was that possible? Find out here!

A long, long journey

Earth's continents are very different today than they were a billion years ago. Different sizes, and shapes, and positions.

Why? Because the continents are moving! They – and the oceans – are carried on the **plates** you met on page 13. As the plates move around they can break up, or join to other plates. So the continents and oceans change too.

From 550 million years ago

Let's start from 550 million years ago (550 **mya**), to see how the continents have changed. Look at their names. And look out for the land shown in red. It's the land that forms the British Isles today!

Did you know?
- The plate you live on is moving eastwards at about 1 cm a year.

What if ...
... the British Isles moved back to the Equator?

Did you know?
- Using GPS, we can track plate movements very precisely.

1

550 mya. The land that will become the British Isles is in two places, far apart. Most is quite far south, in Gondwana. The rest is in Laurentia.

2

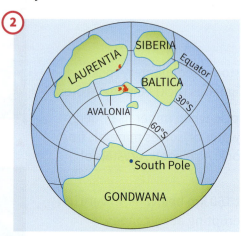

450 mya. A mini continent, Avalonia, broke away from Gondwana some time ago, and is now heading north. Look where the red patches are.

3

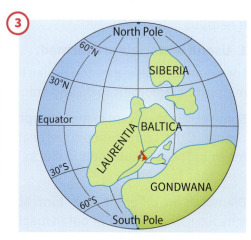

395 mya. Avalonia and Baltica have collided with Laurentia. The land that will form the British Isles is together at last, near the tropics.

4

300 mya. By now all of Earth's land has joined to form a super-continent, **Pangaea**. And look! Our land has crossed the Equator!

5

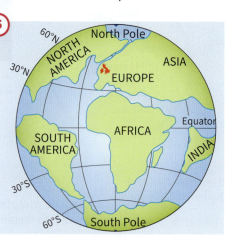

65 mya. Pangaea lasted for about 160 million years. Then it broke up into new continents. Here they are, 65 mya. Do any look familiar?

6

Are we there yet?

Today. The continents as they are now. They are still moving, very slowly. If you come back in say 100 million years, the globe will look different.

What about living things?

What life did the continents carry, as they moved and changed? The **geological timescale** on the right will help to answer that question.

It shows that 550 mya (globe ①), Earth was in the **Precambrian eon**. There were sponges and other simple soft-bodied animals. But nothing with shells or bones yet, and no life on land.

By 450 mya, land plants had appeared. (Globe ②.)

By 200 mya, on Pangaea, there were dinosaurs, reptiles, lush forests, and more.

By 65 mya, the continents looked more like today's. And the age of mammals had begun. But no sign of us! The first human species did not appear until 2.6 mya.

Fossils in the UK

As it travelled, the land that's now the UK was sometimes under ice, and sometimes under warm tropical seas. Sometimes it was desert. At each stage, it was home to different species.

That's why the UK has so many fossils from other environments. They tell us about its epic journey.

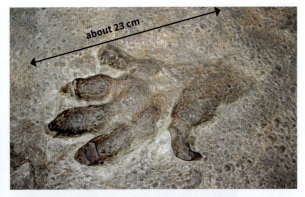
about 23 cm

▲ *A footprint of a reptile, found in Cheshire, UK. This trace fossil, preserved in sandstone, is about 220 million years old.*

The geological timescale

THE PHANEROZOIC EON (OURS)			How long ago?
Era		Period	
			— today
Cenozoic (recent life)		**Quaternary** we (*Homo sapiens*) appear and spread	
			— 2.6 mya
		Neogene apes, chimpanzees, rhinos, horses, sheep …	
			— 23 mya
		Paleogene mammals and birds flourish	
			— 66 mya
Mesozoic (middle life)		**Cretaceous** dinosaurs rule; period ends with their extinction	
			— 145 mya
		Jurassic more dinosaurs appear; first birds	
			— 200 mya
		Triassic first dinosaurs and mammals	
			— 250 mya
Paleozoic (ancient life)		**Permian** first conifer trees; warm-blooded reptiles	
			— 290 mya
		Carboniferous on land: lush forests, reptiles, giant insects	
			— 300 mya
		Devonian first animals on land	
			— 420 mya
		Silurian first bony fish; more land plants	
			— 445 mya
		Ordovician first land plants	
			— 485 mya
		Cambrian first animals with shells appear in the sea	
			— 540 mya
THE PRECAMBRIAN EON first soft-bodied animals appear in the sea			— 600 mya
first living cells appear in the sea			— 3.5 bya
Earth is formed			4.5 bya

Your turn

1 The continents of 550 mya no longer exist today. Why not? Explain in four lines. (Include the word *plates*!)

2 Give *two* differences between the continents of 65 mya and today.

3 Describe the northward journey of the land that is now the British Isles, from 550 mya. (Bullet points, or a flow chart?) Add comments about how its climate may have changed.

4 Look at the geological timescale above. Where did life begin on Earth: on land, or in the sea?

5 Which appeared first on the land: plants or animals?

6 In which geological period did the first birds appear? Give its name, and dates.

7 Many dinosaur fossils have been found in the UK. But none are *less than* 66 million years old. Explain why.

8 Look at the photo of the reptile footprint from Cheshire.

  a In which geological period was this footprint formed?

  b On which continent was it formed?

  c How does it compare with your footprint?

Why has the UK got so many mountains and hills – and different rock types? Find out here.

The relief of the British Isles

Image **A** shows the **relief** of the British Isles – mountains, hills, and low flat land.

The UK has lots of mountains and hills. Why? Because of two key factors: **plate movements**, and **rock type**.

Plate movements

As you saw in Unit 1.5, the land that forms the UK has been on a long, long journey.

It came together when the ancient continents of Baltica, Avalonia and Laurentia collided and joined, through plate movements. (Look at globe ③ on page 14.) Later, all the continents joined to form Pangaea.

When plates collide and join, there is enormous pressure along the edges where they meet. So rock gets squashed upwards. Our mountains are the result of this squashing, many millions of years ago.

The rock around the UK

There's soil and clay and stones over most of the UK. But if you shovel them away, you'll reach solid rock or **bedrock**. Map **D** shows the UK's bedrock.

Look at the different types of sedimentary rock. They formed because the land gathered sediment in different environments, on its travels.

Sometimes the land lay under water. Sediments of dead organisms led to limestone or chalk. Sometimes a river carried in tonnes of sand or mud, leading to sandstone or mudstone. Sometimes the land was a desert, and the wind built sand dunes. These got buried and formed sandstone too.

Scientists can work out the ages of the rocks. They all tell a travel story!

The link between rock type and relief

Rock type plays a big part in the relief of the UK. That's because some types of rock **weather** more easily than others. Then the fragments can be eroded, and carried away – by rivers, glaciers, and wind. In general …

- **higher land is a sign that the rock is more resistant to weathering**.
  Most of the rock in Scotland's highlands is metamorphic rock. Much of it is over 500 million years old. That tells us it does not weather easily!

- **low land is a sign that the rock weathers easily.**
  Look at the land around X on **A** and **D**. The rock here is mudstone, less than 160 million years old. It weathers easily. The result? Low flat land.

A

Key

mountains

↓

low flat land

Yorkshire Dales National Park

X

What if …
… the UK were completely flat?

B

▲ Metamorphic rock on the Isle of Lewis in Scotland. (It's at ○ on **A**.) It is around 3 billion years old – the oldest rock in the UK.

C

▲ *The variety of rocks in the UK gives us some stunning scenery. This shows the hamlet of Wharfe in the Yorkshire Dales National Park, and the area around it. See the park outline on **A** and **D**.*

The bedrock here includes mudstone, limestone and sandstone, between 427 and 448 million years old. The mountain in the background is made of different layers, capped by sandstone.

E

white chalk, around 99 million years old

red chalk, around 101 million years old

sandstone, around 108 million years old

▲ *Three layers of sedimentary rock, of different ages, show up at the cliffs at Hunstanton in Norfolk. They are at ○ on **D**.*

The sandstone formed in shallow seas. The chalk layers formed in deep seas. The red chalk is red because it contains iron compounds.

D   **The bedrock of the UK**

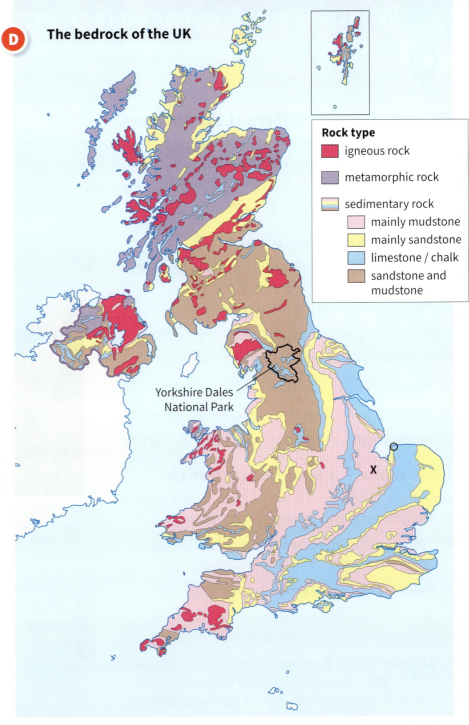

Yorkshire Dales National Park

Rock type
- igneous rock
- metamorphic rock
- sedimentary rock
  - mainly mudstone
  - mainly sandstone
  - limestone / chalk
  - sandstone and mudstone

Your turn

1   Why does the UK have so many mountains?

2   Find dot ○ on image **A**. It's in the mountains.
   a   Give the name of the mountains. (Page 139.)
   b   Identify the main rock type in these mountains.
   c   What evidence is there that this rock resists weathering?

3   Look at photo **B**.
   a   What evidence can you see that the rock here was under great pressure, when continents collided?
   b   Is this rock resistant to weathering? Explain your answer.

4   Map **D** shows the bedrock in the UK.
   a   Define *bedrock*. (Glossary?)
   b   Put the UK's bedrock in order, most common type first:
        metamorphic   igneous   sedimentary

5   Look at photo **C**. The bedrock under the hamlet is mainly mudstone. Which seems to weather more easily, mudstone or sandstone? Give evidence from the photo.

6   **a**   Look at **E**. Why does the top layer have the youngest rock?
   b   In which layer might you find the oldest fossils?

Here you'll explore how landscape around the UK varies with rock type.

Different rocks, different landscapes

The **landscape** is the visible features of an area – natural features such as hills and rivers, and human-built features too.

Landscape varies with rock type. That's mainly because different rocks weather in different ways, and at different rates. Let's look at three examples.

▲ *Where photos A – D were taken.*

Granite – an igneous rock

Granite is resistant to weathering. It breaks down very slowly to sand and clay and stones. It's also **impermeable**: it does not let water soak through.

So where there is granite, you can expect to find:

- land that's higher than the surrounding areas
- thin soil
- boggy areas, because rain can't soak away easily
- sheep farming, since the soil is not good for crops.

Limestone – a sedimentary rock

Limestone is weathered mainly by rain. Rain is slightly acidic, and can slowly dissolve limestone.

- The rain soaks down between blocks of limestone, so gaps get wider as the rock dissolves.
- **Potholes** and **underground caves** form.
- Streams may disappear down **sinkholes** in the limestone, and then emerge somewhere else.
- Since limestone dissolves, the soil in limestone areas is thin, and often stony. It can be difficult to grow crops. So the land is often used for sheep farming.

▼ *The limestone landscape around the village of Malham in the Yorkshire Dales National Park. The bedrock here is around 330 million years old. Note the limestone walls in the fields.*

▲ *In Dartmoor National Park, in Devon, most of the bedrock is granite – about 295 million years old. You can see it at outcrops called **tors**. These formed when some granite weathered more slowly than the rest. The tor shown here is called High Willhays.*

▼ *About 1 km from Malham is this limestone feature, called a **pavement**. The gaps are where rainwater dissolved the rock.*

When's lunch?

The slabs are called **clints**.

The gaps between them are called **grikes**.

Mudstone – a sedimentary rock

Mudstone weathers quickly, to produce clay.

- So you get low flat land, and thick soil.

- Clay swells when it gets wet, and this stops further water soaking through it. So you'll find many ponds and streams.

- Clay is full of the nutrients that plants need. So you'll find plenty of crops grown, where the bedrock is mudstone.

Did you know?
- Around the UK, cliffs made of mudstone are at risk of collapse.

▶ *Farmland in Wiltshire, in an area where the bedrock is mudstone.*

Your turn

1 a Define the term *landscape*.
  b Describe the landscape in photo **B**, in five lines. (Is it flat? Rocky? Broken up? Any sign of human activity?)

2 Photo **A** shows the highest part of Dartmoor – about 620 metres above sea level.
  a Name the bedrock here.
  b Why would this area not be good for growing crops? (You might find some extra evidence in the photo.)

3 a Name the exposed rock structure in: **i A**   **ii C**
  b State *two* ways in which the two structures differ.

4 List the three rock types in this unit in order of resistance to weathering, with the most resistant first. The text will help.

5 Can rock type influence how people earn a living? Explain.

6 Look at the OS map extracts **X**, **Y** and **Z**.
They are for granite, limestone, and mudstone areas.
Match each OS extract to the correct rock type. For each match, give evidence from the OS map.

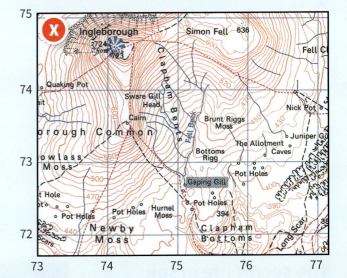

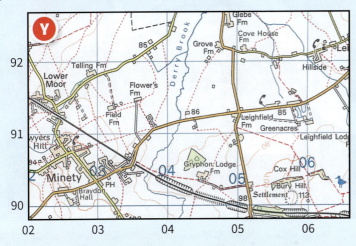

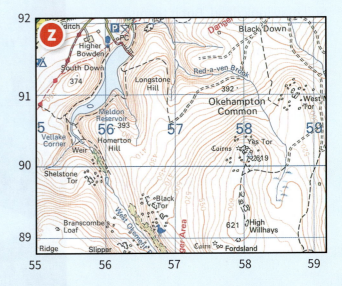

Scale for OS maps 1: 50 000

© Crown copyright

Over time, rock breaks down to give soil. What exactly is soil? And why is it so important to *you*? Find out here.

What is soil?

Soil is a mixture of clay, sand, and rotting vegetation. The clay and sand form when rock is broken down by chemical weathering. (See page 11.)

The soil profile

If you sliced down through soil, you'd see these layers:

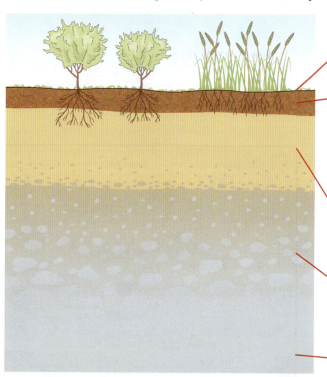

humus. It is a thin layer of rotting vegetation – such as grass and leaves. The **nutrients** in it return to the soil.

topsoil. This layer is rich in humus, and the minerals from the rock. So it is good for growing crops.

The plants take in nutrients such as potassium, phosphorus, calcium, and silicon, which came from the rock – as well as nitrogen, which bacteria in the soil 'fix' from the air.

The sand helps to make topsoil crumbly. So roots can spread easily.

subsoil. It has only a little humus, but it's rich in minerals. Tree roots reach this layer.

rock that is being weathered. It has been broken into chunks already. There are few signs of life this far down.

bedrock. Solid rock, not yet weathered. But it will be, one day.

Did you know?
- Trees 'talk' to each other through a network of fungi in the soil.
- They can pass nutrients to each other along the network.

The secret life in soil

Soil may look dead. But it teems with life: bacteria, fungi, and insects and other animals. It's the most **biodiverse** environment on Earth!

A teaspoon of soil may contain over a billion bacteria of different types. Many feed on dead plant material, 'rotting' it to humus.

Earthworms improve soil. They digest leaves and clay, so their waste is rich in nutrients. And the tunnels they bore allow air and water to circulate.

Yummy!

▶ *Your best friend?*

Soil and you

You may think soil is dirty, and boring, and nothing to do with you. Think again!

Let's start with the parent rock. Its minerals may contain atoms of many different elements.

When the rock is broken down by weathering, and soil forms, the soil also contains those elements.

Yummy.

Plants take in the elements they need through their roots, along with nitrogen and water.

You use them to build up your bones, flesh, teeth, hair, and blood. You'd *die* without them.

Yummy!

You eat plants too – and perhaps animals, and dairy products. That's how elements from rock reach you.

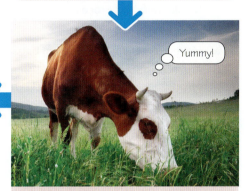

Yummy!

They use them, with carbon dioxide from the air, to make their food. Then animals eat the plants for food.

Looking after soil

There are nearly 8 billion humans on Earth. We depend on soil for most food.

Each crop removes nutrients from the soil. After many crops, soil may be low in nutrients. Further crops won't grow well. In the end, the soil may be useless.

That is why farmers add **fertilisers** to soil. These may be artificial fertilisers, made in factories. Or natural fertilisers such as animal manure.

In many places around the world, the soil is already useless. People may not be able to afford fertiliser. So crops are poor. People may be malnourished.

▲ *Spreading a slurry of animal manure on the land. It adds nutrients to the soil.*

Your turn

1 a What is *soil*?
   b What is the link between soil and rock?
   c Some soils are more *fertile* than others, depending on the bedrock. Suggest a reason. (Glossary?)

2 What is *humus*, and how does it help soil?

3 Which layer of soil is the main one, for growing crops?

4 Explain why soil is an essential resource for humans.

5 a Give one way in which soil may become useless.
   b Fertilisers restore nutrients to soil. What are *fertilisers*?
   c For growing organic food, only natural fertilisers are used. Name one natural fertiliser.

6 Could we run out of soil one day? Discuss!

1 From rock to soil

How much have you learned about rock and soil? Let's see.

check ✓

1 Photo **A** shows an abandoned limestone quarry in Poland.

   a Limestone is formed on the ocean floor. Explain how.

   b Name another rock which is a form of limestone, made from very tiny organisms.

   c Limestone is quarried all over the world, for different uses. Give one use for this rock.

   d The exposed limestone in **A** is being weathered. Define *weathering*.

   e Which type of weathering occurs when rain falls on limestone: *physical* or *chemical*? Explain your answer.

   f Name two geographical features which may form in a limestone area, through weathering by rain.

   g The limestone in **A** is being weathered by another process too. Identify this process, and explain why it can break rock up.

2 a Make a larger copy of diagram **B**. Use at least half a page.

   b Add these labels correctly at **i**, **ii**, and **iii** on your diagram.
   igneous rock   sedimentary rock   metamorphic rock

   c Add these labels to the correct arrows. Curve them?
   *uplifted   uplifted   melts to magma, then cools
   buried under pressure   uplifted*

   d Limestone forms deep in the ocean, at position **i** in the rock cycle. Explain how it can end up on land, as in **A**. Include the term *plates* in your answer.

   e Name the rock that will form at **ii** if limestone is buried.

   f **i** Fossils are found in sedimentary rock. Explain why.

   ii Many fossils found in the UK are of animals that lived around the Equator. Explain why they are now found so far north of the Equator.

3 **C** shows lunch – thanks to rock!

   a The plate and bowl are made of baked clay. How does clay form? (Page 11.)

   b The fork is steel, made from an iron ore mined from Earth's crust. What is an *ore*?

   c Sand is the main ingredient in making glass.

   i What is a *mineral*?   **ii** Name the main mineral in sand.

   d The lunch is pasta with tomatoes and herbs. Pasta is made from wheat. Describe how rock is linked to this food.

   e Earthworms help to make food more nutritious. Explain why.

4 Earth has rock of all ages.

   a The oldest rock found on Earth is around 4 billion years old. It is not plentiful. Suggest a reason.

   b **i** Where could you find rock that is only a day old? (Page 8 has a clue.)

   ii Which type of rock will it be: igneous, sedimentary, or metamorphic?

A

B

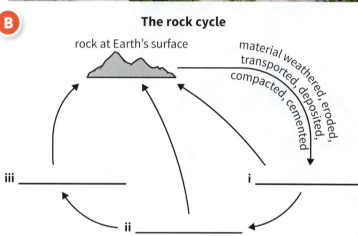

The rock cycle

rock at Earth's surface

material weathered, eroded, transported, deposited, compacted, cemented

iii _____

i _____

ii _____

C

5 a Diamond is a *gemstone*. Name two other gemstones.

   b How are diamonds formed?

   c Suggest a reason why gemstones are often very expensive.

6 *The rock cycle is essential to the way we live.*
To what extent do you agree with the statement in italics?
Write at least ten lines in your answer.

2 Using Earth's resources

We depend on Earth's natural resources. But what are they? And where? And are there enough? Find out more here!

Meeting our needs

Every day, we humans depend on Earth's natural resources.

We arrive on Earth with nothing. Not even clothing. And very soon, we start to rely on Earth for our needs.

We have five basic needs which must be met, if we are to survive. They are shown above.

Our early ancestors made use of the natural resources they found around them, to meet these needs.

About 12 000 years ago we learned how to farm. From then on, soil was an important resource for us.

As time went by, we found more and more ways to use what Earth provides: things like metals, gas, and oil.

Today, we still have the same basic *needs* as our ancestors. There are also lots of things we *want*.

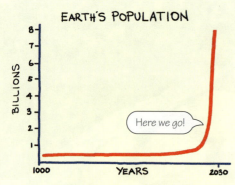

And there's one huge challenge. Our population keeps rising. It has more than doubled in the last 50 years.

So we are using more and more of what Earth provides. But the supply of most natural resources is **finite**!

We already compete with each other for some resources. Some will become scarce. What will we do then?

So what *are* natural resources?

A **natural resource** is something which occurs naturally – without any help from us – and which we can make use of. Like soil, water, wind, sunlight, coal, oil.

We use a lot of plastic. But it is *not* a natural resource. It is made in factories, from chemicals obtained from oil.

Some are renewable

A **renewable resource** is one that we can keep using, and it will not run out. For example, we need sunlight for growing crops. The sun shines on Earth every day, and it won't run out. (Or at least not for another 5 billion years.)

The metal copper is a **non-renewable resource**. It is used in electrical wiring. There is only a finite amount of it in Earth's crust. Millions of tonnes are mined each year, but no more will form. One day, it may be hard to find new deposits.

Where are the natural resources?

Natural resources are found everywhere. But not shared equally!

For example, all countries get rain – but some get very little, and others get lots. And all countries get sunlight, but it is stronger in some than in others.

Our early ancestors made do with the resources around them. But not us! We use resources from all over the world. When you eat an orange, you are making use of the soil, water and sunshine in Spain, or Brazil, or California.

The uneven distribution of some resources, such as oil, has helped to make some countries very wealthy. It has also led to conflict.

In this chapter …

In this chapter we will look at three things we can't live without – water, food, and a source of energy. You will learn about the challenges we humans face in meeting our need for these, with a rising population.

▲ *An electric car contains over 50 kg of copper. A wind turbine to power 500 homes may use over 3 tonnes, including the cables. Demand for copper is expected to soar.*

▲ *Luckily, copper can be recycled over and over, with no loss in quality.*

Your turn

1 What is: **a** a *natural* resource?   **b** a *finite* resource?
   (Glossary?)

2 Think about each item **A – H** listed below.
   a Is it a natural resource? Decide *yes* or *no*, and explain.
   b If your answer is *yes*, give an example of how we use it.

   A   air   B   paper
   C   rock   D   rain
   E   the ocean   F   glass
   G   rainforest   H   chocolate

3 What is:
   a a *renewable* resource?   **b** a *non-renewable* resource?

4 Is this natural resource renewable, or non-renewable? Explain your answer.
   a wind   **b** coal   **c** gold   **d** rain

5 To survive, you need food, water, clothing, shelter, and a source of heat energy (to warm yourself, and / or cook food).
   Now imagine you got shipwrecked in the Pacific Ocean. And you are alone on a small island. You have a bag with two empty glass bottles, a big sharp knife, and a reel of strong wire.
   a Describe your island.
   b Now say how you'll meet those five basic needs. Be creative!

You have already used one amazing natural resource today: water. Does the world have enough of it? Find out here.

There's a lot of water around

Water is a matter of life and death. We can survive only a few days without it.

Earth has a vast amount of water. But most is in the oceans, and it is **salty**. You cannot drink it. Your kidneys would fail. The **fresh** (not salty) water that is available to drink is less than 1% of Earth's total water. Look at **A**:

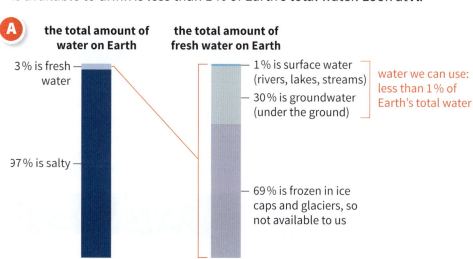

A

the total amount of water on Earth

3% is fresh water

97% is salty

the total amount of fresh water on Earth

1% is surface water (rivers, lakes, streams)

30% is groundwater (under the ground)

water we can use: less than 1% of Earth's total water

69% is frozen in ice caps and glaciers, so not available to us

▲ *Many farms in the UK have boreholes like this one, where water is pumped up from an aquifer to irrigate crops.*

Where do we get our fresh water?

In the UK, around 28% of our total water supply is **groundwater**, pumped up from **aquifers**. The rest is surface water, pumped from rivers, or from **reservoirs** – artificial or natural lakes, fed by rivers and streams.

The pattern is different everywhere. But overall, around a third of the world's population depends on groundwater for everything.

What do we use it for?

This shows our global use of fresh water:

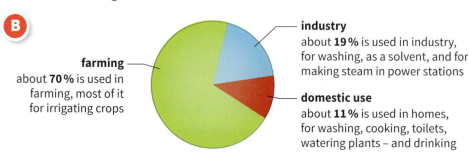

B

farming
about **70%** is used in farming, most of it for irrigating crops

industry
about **19%** is used in industry, for washing, as a solvent, and for making steam in power stations

domestic use
about **11%** is used in homes, for washing, cooking, toilets, watering plants – and drinking

But again the pattern varies around the world:

- Some countries have little industry. Some have little farming.
- Around 3 in 10 of the world's population do not have a water supply at home. Some walk a long way to fetch water from rivers or wells. So they use water very sparingly at home.

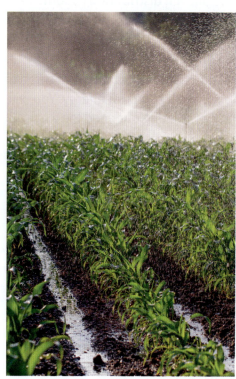

▲ *A sprinkler system irrigating maize, in South Africa. The whole field gets soaked. What happens to the extra water?*

So … do we have enough fresh water?

The answer is **no** – not if we carry on as we are! Many countries have a water crisis already, or soon will. Follow the numbers in **C** to find out why.

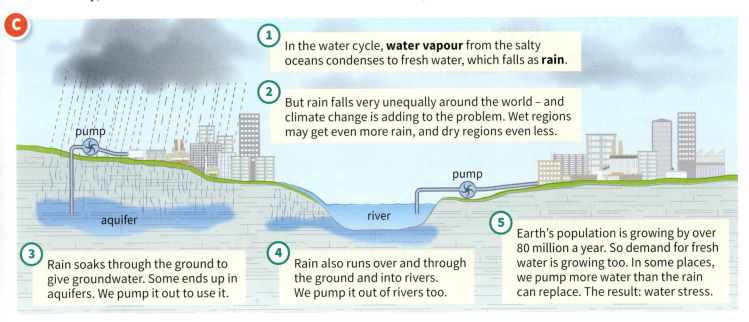

C

1. In the water cycle, **water vapour** from the salty oceans condenses to fresh water, which falls as **rain**.

2. But rain falls very unequally around the world – and climate change is adding to the problem. Wet regions may get even more rain, and dry regions even less.

pump

pump

aquifer

river

3. Rain soaks through the ground to give groundwater. Some ends up in aquifers. We pump it out to use it.

4. Rain also runs over and through the ground and into rivers. We pump it out of rivers too.

5. Earth's population is growing by over 80 million a year. So demand for fresh water is growing too. In some places, we pump more water than the rain can replace. The result: water stress.

Water stress

Water stress is where a country can't meet its demand for fresh water.

Map **D** shows predictions for 2040.

Look at Saudi Arabia. This dry country depends heavily on a big aquifer, which is being depleted.

Brazil has around 6 times as many people as Saudi Arabia – but up to 40 times more rain. Water supply will not be a problem for Brazil, *overall*.

But even where a country has plenty of water, *some* areas may suffer water stress. People, rainfall, rivers, and aquifers, are not spread evenly! So some parts of Brazil face water shortages from time to time.

D

Water stress around the world, 2040

SAUDI ARABIA

BRAZIL

Level of water stress
- low
- low to medium
- medium to high
- high
- extremely high

▲ The level of water stress is calculated by comparing how much water is extracted (pumped out) to how much is available.

6 Using **D**, and pages 140–141, name four countries:

Your turn

1 Nature turns *salt water* from the ocean into *fresh water* that we humans can drink. Explain how this happens.

2 Of all the water on our planet, about what % can we drink?

3 a Define: **i** *groundwater*   **ii** *aquifer*   (Glossary?)

   b What difficulties might there be in getting water from an aquifer? Suggest two.

4 *Most of the fresh water humans use is for drinking.*
   True or false? If false, write sentences to correct this statement.

5 a What is *water stress*?

   b From **C**, give *two* reasons why water stress is increasing.

6 Using **D**, and pages 140–141, name four countries:

   a that may suffer extremely high water stress by 2040

   b that can expect only low water stress by 2040

7 a What level of water stress is predicted for the UK, by 2040?

   b Water stress will vary across the UK. The south east is likely to experience the most. Suggest two reasons for this.

More and more water is being pumped from aquifers around the world. Is this a problem? Find out here.

The end of the good days?

Doug is 85 today. He looks out over the farm. The sprinkler is sprinkling. The corn is growing. He should be happy. But he's not.

He thinks back over his life. Things were so different in Kansas when he was a boy. Farming was tough. Droughts were common. Soil dried up. Crops failed. He still remembers the dust storms. It was hard to survive.

Then came World War II. After it ended, the good days began. You could buy motor-driven pumps, to pump water from the Ogallala. Endless water. Drought was no longer a problem. His father bought more and more land. They grew wheat and corn and soybean.

Now Brad, his son, runs the farm. He wants Brad and the kids to have a good life. But he is worried. The good days may be coming to an end … because the Ogallala is running out of water.

What is the Ogallala?

The Ogallala is a huge aquifer. It covers 450 000 sq km – more than twice the area of Britain. It lies beneath the Great Plains, a region which runs from Canada down the middle of the USA. Eight American states share the aquifer. Look at **A** and **B**.

Most of the water in the Ogallala is at least 8000 years old. So it is called **fossil water**. It has been trapped there since the last ice age.

▲ The Ogallala aquifer lies below the Great Plains, which extend into Canada.

▲ These states share the Ogallala.

▲ Using the fossil water from the Ogallala to irrigate wheat. The green circles are crops. Giant arms rotate like a clock hand, sprinkling the crops with water. This system is called **centre pivot irrigation**. Look how long the sprinkler arms are, compared to the buildings.

▲ In the USA in the 1930s, the Great Plains were called the Dustbowl. Droughts were frequent. Fierce winds whipped dust from the dry soil. Dust storms even reached New York.

And the problem is …?

There is not much rain in the Ogallala area. So there's not much water trickling down into the aquifer.

But thousands of litres are pumped out of it each day. In Kansas alone, 39 000 pumps suck water from it.

So it is not surprising that the water level is falling. It has dropped more than 90 metres in some places, since pumping began.

Farmers drill ever deeper, with more powerful pumps, to get water. So the problem gets worse.

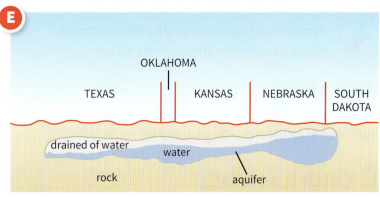

▲ A simplified cross-section through the Ogallala. The aquifer is deepest under Nebraska – so farmers there can rely on it for longer.

So … what will the future bring?

In recent years, farmers have been asked to limit how much they pump from the Ogallala, to make the water last longer. But even if they all stop pumping, it could take 6000 years for the aquifer to refill.

At the present rate, many parts of the Ogallala will run dry within 20 years. Then … disaster. With only rain to water crops, and frequent drought, many farms will fail. Farming communities will be devastated. Towns will be deserted.

People far beyond the Ogallala will feel the effect. Thanks to the aquifer, the area became the **breadbasket** of the USA. It provides a large share of the wheat and other grain crops used in the USA, and for export.

Not only the Ogallala …

There are aquifers of all sizes around the world. Some are much larger than the Ogallala. Many are being depleted.

The **Nubian Sandstone Aquifer** lies under the Sahara Desert. It is over 5 times the size of the Ogallala (2.6 million sq km). It is shared by Egypt, Libya, Sudan, and Chad. See page 141. Libya depends on it most. Water is pumped into a network of pipelines across Libya, called the **Great Man-made River**.

Hardly any rainwater trickles into this aquifer, because it lies in an arid region. It still holds a vast amount of water. But it will not last forever.

Another ruined crop …

▲ Without the water from the Ogallala, farmers must cope with frequent drought.

Your turn

1 a *What* and *where* is the Ogallala? Give a full answer.
  b Name any four states that share the Ogallala.
  c Why is the Ogallala so important to farmers?

2 The water in the Ogallala is *fossil water*.
  a Explain the term in italics. (Glossary?)
  b The level of water in the aquifer is falling. Why?
  c Under which state in **E** is the level lowest, overall?

3 Look at photo **D**. What are those dark clouds made of? And why did they form?

4 Why are there big green circles, in photo **C**?

5 What differences might you expect to see, if you were to visit the place in **C** in 50 years' time?

6 Many groups of people will be affected when the Ogallala can no longer provide water. Name as many as you can. On a spider map?

7 *The water in the Nubian Sandstone Aquifer is fossil water.* True or false? Decide, and explain your logic.

8 a Which countries share the Nubian Sandstone Aquifer?
  b Could sharing an aquifer with other countries lead to conflict? Decide, and justify your decision.

The world faces increasing water stress. What can we do about it? Read on …

What to do?

Water stress is creeping across our world. Let's look at approaches to meeting the demand for fresh water.

1 Catch it as it falls!

Look at this rain: fresh water. Most will run down the drain and off to a river. But what if you catch it first?

More people everywhere are harvesting the rain – from containers on the island of Tuvalu in the Pacific (above), to **harvesting ponds** in Indian villages (below). The pond water is used for crops and animals.

Harvesting ponds fill up when there is heavy rain, for example during India's monsoon season.

2 Stop wasting it!

Water is precious – but we waste a lot at home, in bathrooms and kitchens.

In the UK in 2020, *over 3 billion litres of water a day were also wasted from leaking water mains!* Leaks in the supply are a problem in many countries.

Fixed it!

But around the world, agriculture is the main user of fresh water – and where most waste is.

Instead of soaking the whole field, farmers are trying more careful ways to water crops. In **drip irrigation**, as below, the water runs beside or under the plants, in flexible tubes. Water drips out through tiny holes.

In areas with higher water stress, farmers are also switching to less thirsty crops. For example, from cotton to vegetables.

3 Recycle it!

In the UK and many other countries, waste water from homes flows to sewage works. It is cleaned up, and piped to rivers. It flows off to the sea.

But in some countries, the cleaned-up water is not sent to the river. It is used for farming. In a few cities around the world, it is super-cleaned and used for drinking!

The sewage plant above is in Israel. 90 % of all waste water in Israel is recycled for farming.

4 Move it!

The south of China gets plenty of rain. The north is dry – but it has much of China's agriculture.

China's solution? To move water from the River Yangtze north, in the **South-North Water Transfer Project**.

It aims to transfer 45 billion cubic metres of water a year to the north, carried by canals and tunnels. Completion date: 2050. Cost so far: over $79 billion!

5 Take it from the sea!

Nature turns salty sea water into fresh water, in the water cycle. We can do it faster, in a process called **desalination**.

- Sea water is evaporated and then condensed, leaving the salts behind. OR it is forced through an ultra-fine membrane, which traps them.

- The desalinated water is given further treatment, and then piped into the water supply.

Desalination needs a great deal of electricity. So it is used most in countries with coastlines and plenty of cheap oil or gas for generating electricity. For example, in countries of the Arabian Peninsula (page 124).

But more countries are building **desalination plants**. In the future, these may use electricity from renewable sources (such as solar power).

The UK has one desalination plant so far: on a bank of the Thames Estuary.

Can approaches **1 – 5**, together, avert a severe world water crisis? Time will tell!

Your turn

1. Look at approaches **1 – 5** for tackling water stress. Write a short summary of each, *in your own words*.

2. Of the five approaches to tackling water stress:
   a. which two do you think would cost least? Why?
   b. which could be implemented in the UK? Give examples, or suggestions, for each choice.
   c. which might help a landlocked country with little rain?
   d. which one will impact the environment most? Explain.

3. Why is China willing to spend so much on water transfer?

4. Turn to map **D** on page 27. Name four countries facing extreme water stress, that could make use of desalination.

5. Overall, which approach **1 – 5** might help the world *most*? Explain your choice.

6. Look what this person suggests. Is it a good idea? Explain.

   If there's not enough water in the place ... just move people out!

7. Could **1 – 5** *together* save the world from a severe water crisis? Discuss!

Food, like water, is an essential resource. Is there a risk of the UK running out of food – now or in the future? Find out here.

A

▲ *Delicious fruit and vegetables, from the UK and around the world.*

Are we food-secure?

Hungry? No food in the house? Food is never far away, in the UK. Step into the nearest supermarket, or market, or village shop. Or place an order online.

Overall, the UK has a high level of **food security**.

Food security is when people have access, at all times, to enough safe and nutritious food for a healthy and active life.

But not everyone in the UK can afford enough food. Many families struggle with **food insecurity**. They may need to use a food bank.

It's not the same as self-sufficient!

There is plenty of food available in the UK. But it is not all produced here!

- Overall, around 64% of the food we eat in the UK is produced here. The rest is imported.

- So, although we have a high level of food security, we are far from **self-sufficient** in food.

- A high % of our imported food is grown in warmer climates. Bananas, mangoes, oranges, lemons, limes, ginger, pepper, rice, tea, coffee … We can enjoy these all year round.

Note that we export food too. It earned us around £14 billion in 2019. But we spend far more on imported food: over £39 billion in 2019.

▼ *Overall, the UK has a good climate for farming. Not too hot. Not too cold. Not too damp. Not too dry. About 70% of the land area in the UK is used for farming.*

B

Is our food security at risk?

The answer to this question is *yes*! Look at these reasons:

1 Population growth

- The world's population is rising.
- As countries get better off, people want a wider range of food.
- So there is growing competition for food to import.

2 Politics

- Countries make **trade deals** with each other about all kinds of things, including food.
- For example the UK trades food and other goods freely with the **European Union** of 27 countries – the **EU** – without paying **tariffs** (taxes).
- We obtain around 30% of our food from the EU. Look out for oranges from Spain, and tomatoes from Italy.
- But there may be tariffs on trade with other countries, that affect food imports. And countries may increase tariffs, leading to **trade wars**. (No actual fighting!)

So … we should not rely too much on food imports, for political reasons.

3 Climate change

This is the big one …

- Climate change may mean that other countries have less food to sell us.
- British farmers may suffer too, with more heatwaves, and floods.
- We expect new pests and diseases to appear, as the UK gets warmer. Crops, and farm animals, are at risk.

So … what lies ahead?

We do not know what lies ahead! That is why many experts think the UK must become more self-sufficient in food. Perhaps 80% instead of 55%.

We could build more large **greenhouses**, heated by electricity from renewable sources, to grow food all year round. We could use **hydroponics**. We could grow meat in factories, as in **C**. We will probably not go hungry.

▲ The world's first cultured meat for sale. (Approved by regulators in Singapore, in 2020.) It was 'grown' from chicken cells. No need to slaughter chickens!

▲ Hydroponics. Plants are grown in water containing the nutrients they need. No soil! LED lights can provide artificial sunlight.

Your turn

1 Define these terms about food:
  a *food security* b *food insecurity* c *self-sufficiency*

2 a What is a *food bank*? (Glossary?)
  b When Covid-19 hit the UK, the number of people using food banks surged. Give at least one reason.

3 Suggest two ways the UK benefits from importing food.

4 Explain how these can affect food security in the UK:
  a population growth around the world
  b tariffs on food from other countries

5 Do you agree that climate change could increase food insecurity in the UK? Give reasons.

6 Photos **B** and **D** show two ways to grow crops.
  a Identify three differences between these methods.
  b Name the method used in **D**.
  c Which method would be more suitable for use in a city?

7 a Look at **C**. Would you eat these chicken bites? Why?
  b What are the advantages of producing meat this way? Give as many as you can. Don't forget food security!

Around 2 billion people around the world live with food insecurity. Many go to bed hungry every night. Why?

What's for dinner?

Chifundo looks into her maize store. There is not much maize left. Insects have got in. Some of it looks mouldy. But it has to last another two weeks, before the new crop is ready.

So she'll take some for the evening meal. Maize porridge again, as usual. And only one tomato and one onion, to make a sauce.

She'll sieve the maize kernels to get rid of the dust and insects. She'll grind them into flour. And then she'll boil the flour with water from the well.

They are down to just one meal a day now. The two older children do not complain, but the little one sometimes cries with hunger.

The crop will be poor again this year. The rains stopped early. But it is not just the rains. The soil is worn out. If only she had money for fertilisers. If only she had a new hoe. If only her husband was still alive. If only …

But she will be brave and find a way, for her children's sake.

▲ *Making maize porridge (nsima) on an open fire. This is the staple food in Malawi. It is served with a sauce or, if you can afford it, side dishes of meat and vegetables.*

▲ *Maize kernels, which are ground to make maize flour. (Maize is also called corn. It is related to our corn-on-the-cob.)*

Food insecurity and hunger

Chifundo and her family live in Malawi in East Africa. It is marked on map **A**. She and her family are **food-insecure**. They do not have enough food for a healthy and active life. In fact they go hungry much of the time.

Hunger is a strong indicator of food insecurity. As **A** shows, it is widespread.

A   **Hunger around the world**

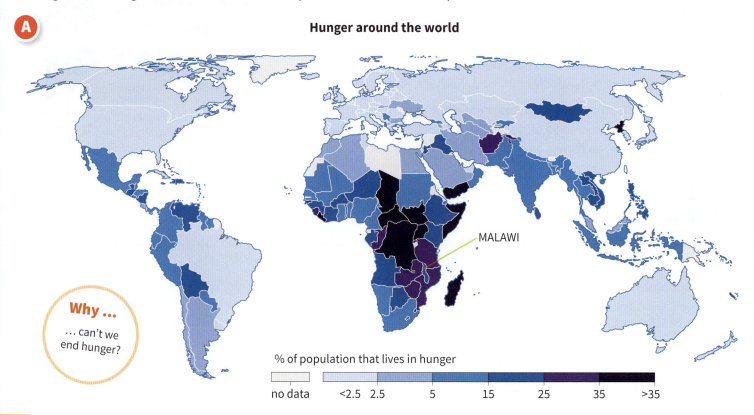

MALAWI

Why …

… can't we end hunger?

% of population that lives in hunger

no data   <2.5 2.5   5   15   25   35   >35

Why is there so much food insecurity?

Every country has some people with food insecurity. But many have large numbers of people living in hunger. Many of these have a high % of farmers!

How can there be food insecurity, where there are lots of farmers? Poverty is a key factor – but not the only one. Look:

1 Poverty in the household
- can't afford to buy enough food

2 Poverty on the farm
- can't afford fertiliser for poor soil
- can't afford good-quality seeds, or pesticides
- can't afford farm equipment

3 Poverty across the country
- poor roads and no electricity make it hard to move food, and store it safely
- little help or training for farmers

4 Pests and diseases
- destroy at least 10 % of all crops around the world each year, and much more in some places

7 Conflict
- farmers may be forced to flee
- movement of food may be blocked

6 Climate change
- rainfall more irregular, drought and floods more frequent
- crops and farm animals suffer
- pests and diseases spread further

5 Water stress
- not enough water in some places to irrigate crops

The impacts of food insecurity

- If people do not have enough to eat, they do not have energy for work. And they are more likely to fall ill. So they end up even poorer.
- Children's development may suffer. This will affect them for life.
- When food is in short supply, its price rises. People can afford even less.
- People try hard to grow more food – so worn-out soil gets even worse.
- Food insecurity can tip over into famine – for example, if drought strikes. People have no food stores to fall back on. The result: a crisis.

▶ *Plants depend on soil for nutrients. When soil loses its nutrients, crops cannot grow well. Fertilisers restore nutrients. Animal manure is a natural fertiliser.*

Your turn

1 Is there a link between *food insecurity* and *hunger*? Explain.

2 a From map **A**, name the continent with the highest % of people living in hunger, overall.
  b Name three countries in that continent with over 35 % of their people living in hunger. Page 141 will help.
  c Comment on the extent of hunger in Malawi.

3 Chifundo's new maize crop will be poor.
  a Give one reason for this, linked to poverty.
  b Give one reason that may be linked to climate change.

4 Explain clearly the link between food insecurity and …
  a water b roads c conflict d soil e crop pests

5 Look at the factors in **B**. In your view, which factor:
  a would be easiest for a government to tackle? Why?
  b can a country *not* tackle alone? Why?

6 Malawi is a peaceful country, so factor **7** in **B** does not apply. Imagine you are Chifundo. Which *two* of the other factors would you want the government to tackle first? Why?

This unit shows some ways to tackle food insecurity – from seeds to satellites!

A better day for Chifundo?

You met Chifundo on page 34. Can her life improve? Let's see.

Joy!

My maize crop was brilliant this year. I have 14 big sacks of kernels! And I grew beans and groundnuts too. I'll be able to sell a lot.

So … what happened? People from an NGO came to the village last year. They gave us seeds and fertiliser. They showed us a better way to plant, and how to dig pits to trap rain. They gave us special sacks for storing the maize kernels, that keep the mould and insects out.

The help is like a loan. I will pay them back when I earn enough. Soon …

Another wonderful thing … the track to the town is a proper road now. So I will wave a truck down at 6 in the morning to bring my produce to the market. No more trudging for hours with heavy sacks.

I will buy two chickens in the market tomorrow, so we can have eggs.

But best of all is this. When I sell my produce I will earn enough to send Teleza, my eldest, to school.

So …with a little help, the food security for Chifundo's family has been improved. Do you agree?

Imagine if this help was given to every farmer like Chifundo. Food security would improve for the whole country – not just for farmers.

▲ Seeds for food security! Breeders cross-breed plants to give improved qualities. For example, greater resistance to drought. The result is hybrid seeds.

▲ Sacks for food security! Lots of maize gets spoiled in storage. These sacks have three layers, and keep insects and oxygen out. The maize can be stored for up to two years.

▲ Fertiliser for food security! Chifundo learned to put a little around each plant, rather than scattering it everywhere.

▲ Rainwater for food security! An expert demonstrates how planting in pits can help to trap and store rainwater.

Meanwhile, in Togo …

Chifundo is feeling happy, on her farm in Malawi. 4000 km away, in Togo, a government worker is feeling happy too.

Food insecurity, technology, and Covid-19

We are a small country. Around 8 million people. Much poverty. When Covid-19 struck, we had lockdowns. Many thousands of people had no work – so no money for food. Food insecurity shot up.

Our government wanted to help the poorest people – fast. But where were they? And how to help them fast? The answer was technology!

First, satellite images
Experts in the USA helped us. They used artificial intelligence to study satellite images of Togo. Metal roofs and well made roads were ignored. Thatched roofs and rough tracks showed the poorest areas.

So now we knew *who* the poorest people were. Because adults in Togo had registered to vote in their areas, a while back.

Then, mobile phones
Even our poorest people in Togo have basic mobiles. So we checked with the phone companies. Then we sent texts to people, asking them to register for help. Within minutes, we sent them vouchers on their mobiles. They could take these to stores and buy food, or get cash.

Success? Oh yes!

▲ *Mobiles for food security! A voucher on his mobile may have helped this villager in Togo with food, at least for a time.*

Technology and food insecurity

Satellites, drones, and mobiles, are all helping to tackle food insecurity. For example, suppose a new disease appears on banana farms.

- Local experts take close-up photos of diseased plants, and aerial photos using drones.

- Satellites take images of a much larger area.

- All the images are fed into a platform that uses artificial intelligence. It analyses them, and can detect where, and how fast, the disease is spreading.

- Farmers, and governments, are warned.

▲ *Satellites for food security! Between them, Europe's two Satellite-2 satellites pass over an area every 5 days. Sensors can detect moisture levels in crops in a field, and even poor soil.*

Your turn

1 Chifundo has grown enough food to sell some. Identify ways in which she and her family can benefit from selling food. You could use a spider map.

2 An NGO helped Chifundo and the other farmers.
   a What is an *NGO*? (Glossary?)
   b NGOs rely on donations – including from people like you. Over time, Chifundo must pay the NGO back for what it gave her. Do you think this is fair? Explain.

3 a Satellites help us to tackle food insecurity. Explain how.
   b Artificial intelligence helped to analyse images for Togo. What is *artificial intelligence*? (Glossary?)

4 Imagine you are Chifundo. Create a spider map, with notes, of items and activities that will help you to produce even more food next year. Will you include anything from this page?

5 Think about all the ideas you met in this unit. Could they save the whole world from food insecurity? Discuss!

For nearly 300 years, fossil fuels have provided the world's energy. But are their days numbered? Find out more here.

The world depends on the fossil fuels

You know about the fossil fuels already: **coal**, **oil** and **gas**. How do we use them?

Coal is widely used in power stations around the world for generating electricity, since it is the cheapest fossil fuel. And in making steel.

Oil is king for transport – in the form of petrol, diesel, jet fuel, and more. Fuel oil is burned in some power stations, and home heating systems.

Natural gas is burned in many power stations. It is also piped to homes in many countries – including the UK – as the fuel for heating and cooking.

The fossil fuels are still forming, somewhere on Earth. But we are using them up much faster than they form. That makes them a **non-renewable resource**.

How their use has grown

Graph **A** shows how our use of fossil fuels has grown since 1800, when the Industrial Revolution was underway. **Wood fuel** is *not* a fossil fuel, but it is in for comparison. It includes logs, firewood, and charcoal.

A

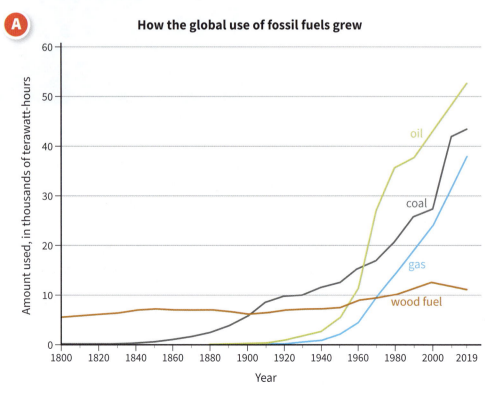

How the global use of fossil fuels grew

(Graph: Amount used, in thousands of terawatt-hours, versus Year from 1800 to 2019. Lines shown for oil, coal, gas, and wood fuel.)

▲ *Bringing home the firewood, in Vietnam. Over 2 billion people in developing countries depend on firewood and charcoal, for cooking and heating. So forests disappear.*

And the bad news ...

Fossil fuels have brought us enormous benefits. But as you know, we are now paying a big price.

Carbon dioxide forms when fossil fuels burn. It is a **greenhouse gas**. Its levels in the atmosphere are rising. So Earth is getting warmer, and climates are changing. This is affecting all life on Earth.

So ... is it goodbye to the fossil fuels?

In 2015, 196 countries signed the **Paris Agreement** to cut greenhouse gas emissions. The aim: to limit the global temperature rise to less than 2°C above pre-industrial levels.

This is the only way to avoid the worst impacts of climate change.

Many countries, including the UK and USA, plan to be **carbon neutral** by 2050. China aims for 2060. It means cutting back hugely on fossil fuels.

B shows how much the world depended on fossil fuels in 2019. Can we be carbon neutral by 2050, or even 2060? It is an enormous challenge.

▲ *Planting trees will help us to become carbon neutral. They take carbon dioxide from the air for photosynthesis.*

B

Where the world got its energy in 2019

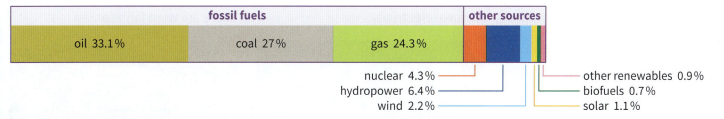

fossil fuels			other sources	
oil 33.1%	coal 27%	gas 24.3%		

nuclear 4.3%
hydropower 6.4%
wind 2.2%

other renewables 0.9%
biofuels 0.7%
solar 1.1%

Renewable energy sources

We still need energy! So the world is switching to electricity from renewable energy sources, to power everything – homes, cars, industry. Find out how this **green electricity** is generated, in the next unit. We take the UK as example.

Your turn

1 We have to *rbnu* fossil fuels to release their *geerny*. Unjumble the two jumbled words!

2 This is about graph **A**.
   a Describe the trend in demand for all three fossil fuels.
   b Which fossil fuel did we use most of, in 2019?
   c Did oil drive the Industrial Revolution? How can you tell?

3 a We obtain different fuels from oil. Name two.
   b The Model-T Ford car was launched in the USA in 1908. It kicked off the age of the car. How did oil consumption change after this, and why? (Check **A**.)
   c The demand for oil fell sharply in 2020 (not shown on **A**). It would rise again later. Suggest a reason for its fall.

4 Wood fuel is also shown on **A**, for comparison.
   a Name three forms of wood fuel.
   b Why do so many people still depend on wood fuel?

5 Most countries signed the *Paris Agreement* in 2015.
   a This was in response to a problem. What problem?
   b The Paris Agreement means that the world must cut back sharply on using fossil fuels. Explain why.
   c In what ways could the Paris Agreement change how *you* live your life?

6 a What does *carbon neutral* mean? (Glossary?)
   b Using the data in **B** (for 2019), comment on the challenge the world faces in going carbon neutral, by say 2060.

Find out how electricity is generated from renewable sources in the UK – and how these sources are linked to our geography!

What if...
... oil and gas had not been discovered?

Electricity from renewable sources

By 2050, most electricity in the UK will be **green**. That means, from **renewable energy sources**. We will use it for heating, lighting, and our electric cars.

First, *check J on page 41* to see how turbines are used in making electricity. Then look at these ways, **A – F**, that we can make green electricity in the UK.

A Some power stations burn **biomass** (wood, straw …) or **waste**, to boil water, for steam to spin their turbines.

B In a **hydroelectric** power station, a fast-flowing river spins the turbines to make electricity.

C On a **wind farm**, wind turns the blades. They are joined to the turbine. There are wind farms at sea too.

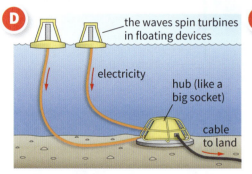

D the waves spin turbines in floating devices / electricity / hub (like a big socket) / cable to land

With sea all around us, engineers are trying out ways to make **waves** spin turbines. They hope for success.

E As it rises and falls, the **tide** can also spin turbines. Several are installed off the coast of Scotland.

F But on **solar farms** like this one in Oxfordshire: no turbines! Sunlight strikes **PV cells**, giving electricity.

Next, study the other graphics on page 41. Then do *Your turn*.

Your turn

1 With the help of **J**, draw a flow chart to show how electricity is generated using a turbine. Add a title.

2 Define: **a** *biomass* **b** *PV cell* (Glossary?)

3 **A – F** shows six ways to generate electricity. Which of them:
   a produce no harmful gases? **b** do not need turbines?

4 Map **H** shows the location of solar farms in the UK.
   a First, *describe* the pattern in their location.
   b Now *explain* it. **G** will help. Include the word *Equator*!

5 Is there a link between wind speeds (map **I**) and the location of land-based wind farms on **H**? *Explain* any link you find.

6 Compare **H** with the map on page 139. Is there a link between land height and the location of hydroelectric stations? *Describe* and *explain* any link you find.

7 Look at pie chart **K**, for electricity in the UK in 2019.
   a Identify the main source for electricity that year.
   b Calculate the total % we got from renewables.
   c Can we be carbon neutral by 2050? Give your opinion!

8 **a** Which of the UK's renewable energy sources depend on:
      i our location on Earth? **ii** our physical geography?
   b Name a country with fewer sources of renewable energy than the UK has. Explain your choice.

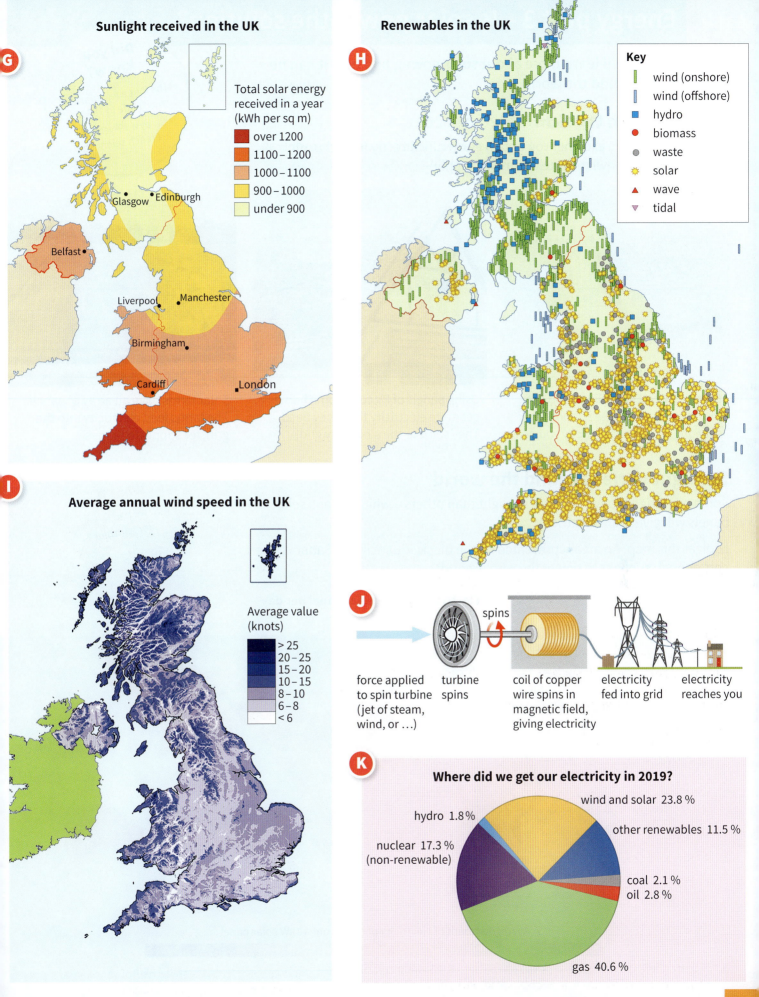

Sunlight received in the UK

G

Total solar energy received in a year (kWh per sq m)

- over 1200
- 1100 – 1200
- 1000 – 1100
- 900 – 1000
- under 900

Glasgow • Edinburgh
Belfast •
Liverpool • Manchester
Birmingham
Cardiff • London

Renewables in the UK

H

Key
- wind (onshore)
- wind (offshore)
- hydro
- biomass
- waste
- solar
- wave
- tidal

Average annual wind speed in the UK

I

Average value (knots)
- > 25
- 20 – 25
- 15 – 20
- 10 – 15
- 8 – 10
- 6 – 8
- < 6

J

spins

force applied to spin turbine (jet of steam, wind, or …) | turbine spins | coil of copper wire spins in magnetic field, giving electricity | electricity fed into grid | electricity reaches you

K

Where did we get our electricity in 2019?

- wind and solar 23.8 %
- other renewables 11.5 %
- hydro 1.8 %
- nuclear 17.3 % (non-renewable)
- coal 2.1 %
- oil 2.8 %
- gas 40.6 %

Here you'll learn more about solar power, and how it can be used around the world.

Straight from the Sun

Solar power is very exciting. You obtain electricity directly from the Sun. You can have your own private electricity supply!

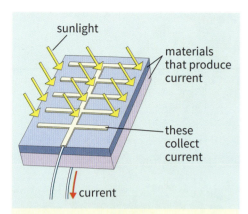

Place a **PV cell** in sunlight. You get instant electricity! *The stronger the sunlight, the more electricity you get.*

Using panels of PV cells, some homes get more electricity than they need. They sell it to the electricity grid!

Solar farms have an array of panels. The farms feed electricity into the grid, for supply to homes.

Sunlight varies around the world

Some countries receive stronger sunlight than others – which means solar panels will give more electricity.

Look at this map. An area represented by the black square in the Sahara Desert gets enough sunlight to power the whole world!

Electricity from a solar panel per day

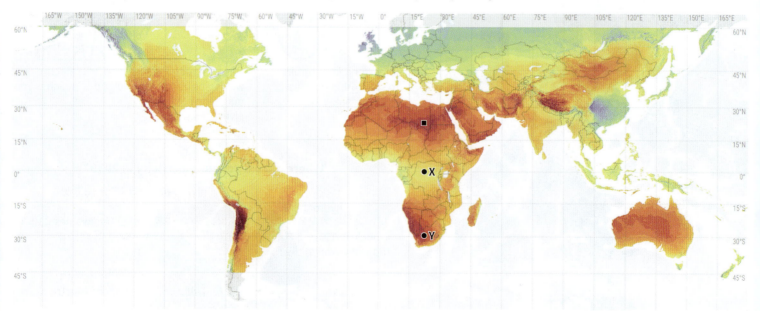

Average amount of electricity per day from a 1kW solar panel

2.0 ———— increasing ————→ 6.4 kWh

▲ *Nomadic herders at their ger (home) on the steppe in Mongolia. When they move, the solar panels and satellite dish are packed up too.*

▲ *Off to school in Côte d'Ivoire. Solar panels in their backpacks generate enough electricity to power lamps in the evening for study.*

Is solar power the solution?

Many of the world's poorest countries are in Africa. Many buy in fossil fuels, especially oil for transport – expensive, and contributing to climate change.

But many of them have free strong sunlight! Solar farms would provide clean energy. Solar power can help them to develop, and climb out of poverty.

No need to wait for the grid …

For many people, electricity is still a distant dream.

▲ *Solar power for safe water. River water is poured into the top container, where it's boiled by solar power. (The lid has PV cells.) The steam condenses to give clean water in the lower container.*

- Many of the world's poorest countries do not have enough power stations, and other infrastructure, to extend the electricity grid to everyone.

- By 2019, around 1 in 10 people still had no electricity at home.

- But solar panels are getting cheaper.

- And solar power can be used at any scale. Get electricity just for your home, or the village, or the town, or the whole country!

- So electricity is within reach of everyone. And it transforms lives.

Your turn

1 What is needed, to generate electricity from sunlight?

2 Use the maps on pages 42 and 140 – 141 for this question.
   a Which of these continents gets stronger sunlight, overall?
      Europe   South America   Africa
   b How did you decide your answer to **a**?
   c In which of these will a solar panel give most electricity?
      UK   Spain   Malawi   Chile
   d Explain why the same panel will give more electricity in Spain than in the UK, in a day.

3 Find X and Y on the map. Suggest a reason why the sunlight is stronger at Y, even though X is on the Equator.
   (A jumbled clue: *yoducl*.)

4 a In your opinion, what are the two main advantages of solar power? Write a sentence about each.
   b Does solar power have any *disadvantages*? State them.

5 Look at the two girls in the photo, going off to school. They have no electricity at home. Explain how solar power may transform their lives.

6 Vaccines must be kept cool. Design a mobile unit for carrying out vaccinations in poor rural villages that have no electricity. Your transport: a motorbike.

7 Look back at the renewable sources of energy on page 40. In your opinion, which two can help most, to turn the world carbon neutral by 2060? Give reasons for your choice.

We are putting more and more pressure on Earth's resources – and destroying other species as we go. What to do?

We have only one Earth

We have only one Earth. We depend on its resources, to survive. But look:

- Our population continues to grow. So, every year, we *need* more resources. More fresh water. More food. More energy. More land to live on.

- As countries develop, we also *want* more of everything: cars, mobiles, holidays, clothes, and much, much more. These use up resources too.

- So pressure on Earth's resources is growing – and most are finite.

- And meanwhile, we are causing climate change.

What about other living things?

We share Earth with millions of other species. We have identified around 1.2 million. Scientists think there are at least six times more – hidden in the sea, below ground, and in remote areas on land.

While we seek more resources – including land to live and farm on – we crush other species. Many have been driven to extinction, or close. For example:

▲ *The rainforest is home to many thousands of species. So when it is cleared for farming and other uses …*

There are about 90 Amur leopards left in the wild, in the forests of Russia and China. Their **habitats** have been destroyed by logging, road building, and other developments.

Spix's macaws are probably extinct in the wild in Brazil, since 2000. Their woodland habitats were destroyed. Many were trafficked. Around 160 remain – in captivity.

Qiqi, China's last Yangtze River dolphin, died in 2002 in captivity. The others may have been killed off by accident – tangled in the gear of big fishing boats.

The sixth mass extinction?

From fossils, scientists have identified five big **mass extinctions** in Earth's history. The last was 66 million years ago, when dinosaurs were wiped out. Now another mass extinction is underway. But *this one is caused by us*.

A United Nations report in 2019 sounded the alarm. It said that nearly a million species could be extinct within decades, if we carry on as now.

This loss of **biodiversity** is not only in faraway places. In the UK, 41 % of studied species are in decline.

▶ *Tree sparrows declined by 96 % in the UK between 1970 and 2020. Why? Fewer insects and seeds to eat, and fewer old trees and hedges to nest in.*

▲ Hedgehogs have declined by 95% in the UK since the 1950s. Through loss of habitat, and being run over by cars.

▲ Bumble bees pollinate crops, flowers, and trees. Their decline in the UK is due to pesticide use and other farming practices.

Living sustainably

It is clear by now that we humans must learn to **live in a sustainable way**, if we are to leave this planet fit for future children. Living sustainably covers many areas, including:

- putting an end to hunger, and achieving food security for everyone.
- working urgently to limit climate change
- buying and using things in a responsible way, *so as not to waste Earth's resources*; for example, not buying clothes you will soon throw out
- protecting other species; for example, stopping the destruction of rainforests, and harmful farming practices.

193 countries have agreed to a set of **Sustainable Development Goals**, that cover these issues, and more. The aim: to achieve them by 2030. Can we?

What about you?

You are a global citizen. You can help the world to live sustainably. For example, by not wasting things – including water, food, and electricity. By recycling. By buying Fair Trade foods. By passing the message on.

▲ Rainforest was destroyed to make way for this oil palm plantation in Indonesia. So thousands of species lost their habitat. Palm oil is used in bread, cakes, sweets, soap, shampoo, and many other products.

Your turn

1 Define these terms. (Glossary?)
   a *habitat*
   b *mass extinction*
   c *living sustainably*
   d *biodiversity*

2 There are about 8 times more humans now than in 1800. Summarise the impact of this growth on other species.

3 Look at photo **A**. What do you think has happened to:
   a the animals that used to live here?
   b the plants that used to live here?

4 Write a list of human activities which destroy the habitats of other species. (For example, would you include farming?)

5 Bees are important to our food security. Explain why.

6 You will speak on behalf of one animal from this unit, **B – G**. Choose one (even if extinct). Write down what you will say.

7 Could *you* be linked in any way to the destruction of habitats? Think about it. Then give reasons for your answer.

8 Look at the four bullet points above. Choose any one, and explain why that activity will benefit future generations.

9 Think about these two people's opinions. Then write a response to one of them – or both!

We humans are pests.

Why worry if other species die off? It does not matter!

2 **Using Earth's resources**

How much have you learned about using Earth's resources? Let's see.

check ✓

1 **A** shows locusts arriving on a farm in Africa. There can be millions in a swarm. They can eat their way through fields of crops within hours. They are feared by farmers in many African and Asian countries, and the Middle East.

a Locusts increase food insecurity in a country.

    i Define *food insecurity*.

    ii Explain how locusts increase food insecurity.

b The factors below also play a part in food insecurity. Choose **two**. Outline the part each plays:
poor soil   *water stress*   *poor roads*

c Locusts and other pests are expected to become a bigger problem for farmers in many countries. Why?

d State one way in which the UK increases its food security.

e Name a system in which plants are grown without soil.

A

2 Look at graph **B**, for a recent year. Each dot represents a country, but only some countries are named.

Undernourished means not getting enough food for a healthy life. Look at Malawi. The average wealth per person was $1060. 19% of the population was undernourished.

a Now compare Malawi and Ghana.

    i In which of them were people wealthier, on average?

    ii Which had a lower % undernourished?

b i In which named country in **B** were people wealthiest?

    ii Comment on the undernourishment in that country.

c Graph **B** is called a *sc_____ph*. Complete the word.

d *Overall*, what trend does the graph show? Start your answer like this: *As average wealth per person …*

e Explain the trend you described in **d**.

f Give one impact of undernourishment on a country.

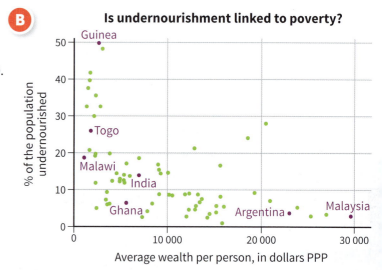

B

Is undernourishment linked to poverty?

% of the population undernourished (vertical axis, 0 to 50)

Average wealth per person, in dollars PPP (horizontal axis, 0 to 30 000)

Countries labelled: Guinea, Togo, Malawi, India, Ghana, Argentina, Malaysia

3 **C** shows one of the UK's **offshore wind farms**.

a i What is a wind farm for?

    ii Wind is a *renewable* source of energy. Explain the term in italics.

    iii Name two other renewable energy sources in the sea.

b New coal-burning power stations are not permitted in the UK, but new wind farms are. Explain this.

c i Is sunlight a renewable source of energy? Explain.

    ii How is electricity obtained from sunlight?

    iii As page 42 shows, African countries are wealthier than the UK in terms of strong sunlight. Why is this?

4 Water stress is increasing around the world.

a Give one key reason for the rise in water stress.

b What is the main use for fresh water, around the world?

c i The water shown in **C** is not fresh. Explain why.

    ii Name the industrial process that could be used to turn the water in **C** into fresh water.

C

5 *We must learn to live sustainably, or our planet will become unfit for future generations.*
Discuss! Consider at least two aspects of our impact on planet Earth. (More if you wish!) Write at least half a page.

Earning a living

Around half of the UK population works for pay. What kind of work do people do? Find out here.

People at work

In the UK, about 33 million people are working for pay. That's a lot of people! And we all depend on them.

First, they provide **goods** and **services** we need.

And second, most pay the government part of what they earn, as **income tax**. This money in turn pays for the National Health Service, and helps to fund schools, the police, and other services.

▲ *Off to work we go.*

What jobs do people do?

There are hundreds of different kinds of jobs. But they fall into four **sectors of employment**. Look:

primary sector

The **primary sector** is where people produce things from the land and sea. For example, farmers, fishermen, miners, oil workers, forestry workers. Often they produce **raw materials** which other people will process.

secondary sector

In the **secondary sector**, people make or build things, often using raw materials from the primary sector. This group includes factory workers, house builders, and people constructing roads and railways.

tertiary sector

In the **tertiary sector**, people provide services for other people. For example teach them, or care for them when they're ill, or sell them things in shops and online, or drive them around in taxis, or entertain them.

quaternary sector

In the **quaternary sector**, people with high-level expertise solve problems, and develop new goods and services, such as vaccines. We treat this small sector as a subset of the tertiary sector, since it offers services.

Where do they work?

Where a person works depends on the sector.

The primary sector

Workers in the primary sector make use of Earth's natural resources. So that dictates where they work. For example, farmers need land for growing crops and rearing animals. So they are based in rural areas.

The secondary sector

In the secondary sector, workers process materials. Factories may set up close to the source of a material, and / or on a good transport route for easy access.

For example, factories that produce frozen vegetables may be in farming areas. So vegetables can be frozen and packed soon after harvesting.

But if you work in construction, you will go where your next job is.

The tertiary sector

Here you provide services for people. Many services can be provided over the internet, so you could work from anywhere. For example, you could do accounts for people online, or offer legal advice.

But if you are a dentist, you must be where people can reach you.

The quaternary sector

Many jobs in this sector depend on **research and development** (or **R & D**) in science and technology. So they may be located in **science parks**. These are often linked to universities. The UK has over 100 science parks.

What is *the economy*?

The economy is all about jobs and money!

It means all the activities going on in a country, in producing, buying, selling, and distributing goods and services. If more goods and services are being produced and sold than before, we say the economy is growing.

▲ *Not moving too far …*

▲ *The economy in action.*

Your turn

1 a What are: **i** *goods*? **ii** *services*? (Glossary?)
  b Give four examples of each.

2 a Name the four employment sectors.
  b Which sector includes *manufacturing*?

3 Identify the sector that this person works in:
  a a quarry worker, extracting limestone
  b a person developing a new camera for a mobile
  c the person delivering the post
  d a carpenter on a building site
  e the lighting director on a film set
  f a hairdresser

4 Name any two jobs, and identify their sectors:
  a where people need to meet face to face
  b where people move from site to site to work
  c where people are tied to one place
  d which could be done from anywhere

5 You will start a factory to make electric cars. Suggest four factors to consider, in choosing a location for it.

6 a What does the term *the economy* mean?
  b Is this part of the economy? Explain your answer.
    i teaching class, in school
    ii playing in a Premier League football match

The UK's employment structure is very different today than in past centuries. Here you can find out how, and why.

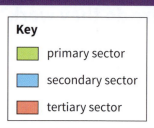
Once upon a time …

Once upon a time, most people were in farming!

1600

Come back!

Baaa.

What do you think?

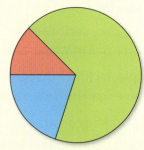

Employment structure in 1600

It's 1600. Unless your family is wealthy, you are probably working hard on a farm already. No school for you! Most people in the UK are in farming.

Other people make things – like shoes, and furniture. Some provide services – for example as merchants, or as servants to richer families.

We can show the **employment structure** for 1600 as a pie chart, like this. The biggest slice is the **primary sector**. Check the colour key.

1850

What has he got now?

It's my new seed drill.

Nine hours a day – and I'm only ten!

Get a move on, laddie!

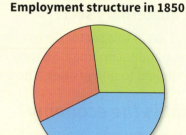

Employment structure in 1850

It's 1850. There have been big changes in farming since 1600. Farms can now produce a lot more food, with fewer people. And there's been…

… an even bigger change: the **Industrial Revolution**. You may be working in a factory now, or a coal mine. (School is still not compulsory.)

Thanks to all the new factories, the **secondary sector** has grown. It will grow further over the next few decades, as more factories open.

1970

Right on, man!

We were once the workshop of the world!

CLOSED

FINE FASHIONS

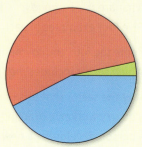

Employment structure in 1970

It's 1970. By now, everyone must go to school until age 15. But meanwhile, out on the streets, life is getting tough for many working people …

… because many factories are closing. They can't compete with factories in other countries, which can make things more cheaply.

Coal mines are closing too. And the % of people in farming is still falling. So the primary sector has shrunk a lot. But look at the **tertiary sector**.

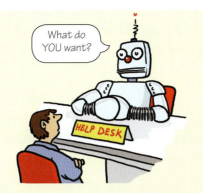

Employment structure today

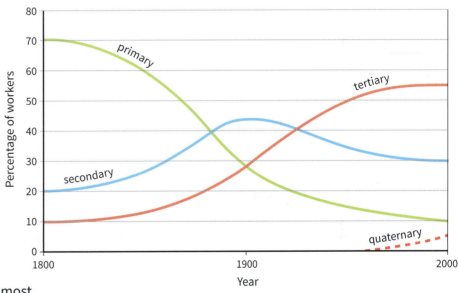

And here we are today. The biggest change since 1970 is computerisation. Computers, mobiles, TVs, washing machines … all have computer chips.

Computers, the internet, and other technologies have led to a rise in new jobs in the service sector – and helped the **quaternary sector** to grow.

So here is today's pie chart. The primary slice has shrunk further. Look how big the tertiary slice is. (It includes the quaternary sector too.)

A graph of the changes

This graph shows how employment structure has changed, in the UK.

It is a **model graph**. That means it aims to show the *pattern* of change, but *not exact numbers*.

Look at the green line. It shows how the % of workers in the primary sector fell, as the UK developed.

Changes are still going on. So the UK's employment structure might be very different, 100 years from now!

Not just the UK

Employment structure has been changing in most countries. But the changes do not always follow the same pattern, or occur at the same speed. You will find out more about this in later chapters.

The changing structure of employment

▲ *This model is called* **the Clark-Fisher model***. When it was proposed, in 1940, the quaternary sector did not exist. Remember, it aims to show the pattern, not exact numbers.*

Your turn

1 Name: **a** three goods **b** three services
we use today, that did not exist in 1850. (Glossary?)

2 Look at the pie chart at the top of this page.
   a What does the symbol ~ beside a number mean?
   b About what % of workers are in the primary sector today?
   c Give two other activities in this sector, besides farming.
   d How has the % of people working in the primary sector changed over the years since 1600?
   e Suggest *two* reasons to explain the changes in **d**.
   f Suppose the primary sector shrank to zero, in the UK. What might the consequences be? Suggest three.

3 Look at the graph above.
   a What does the curved blue line represent?
   b Describe the shape of the blue line.
   c Now *explain* why the blue line has this shape.
   d Could this sector grow again? Explain your answer.

4 The internet became available in 1992. Name two jobs that definitely did not exist before 1992.

5 In 1870, a new law said that children aged 5–13 must attend school. In 1944 the leaving age was raised to 15. Would it be a good idea to leave school at 15 today? Discuss.

Changes in employment affect our lives. This unit is about changes in Doncaster, in South Yorkshire.

Doncaster

Estimated population in 2020
town 111 000
borough 311 000

Doncaster's story

The Romans Doncaster began as a Roman fort at a crossing on the River Don, around 71 CE. It was on the soldiers' route from London to York.

The Middle Ages and beyond By the 13th century, long after the Romans, Doncaster had become a busy market town. By the 16th century, it had become a hub for stage coach travel too. It grew wealthy.

Industrialisation In 1831, the population of Doncaster was about 10 000. Thanks to the Industrial Revolution, it would become a busy industrial centre – mainly because it lay in a coal area. By 1900, coal mining was the biggest employer around Doncaster.

Coal attracted other industries, including steel and glass making. Over 2000 steam trains were built in Doncaster. New railways and canals improved its transport links. By 1959 the population was around 83 000, and growing.

Industrial decline But around 1970, decline set in. Fewer and fewer trains were built in Doncaster. Mines began to close. People lost their jobs. This led to high levels of poverty and ill-health, and areas of deprivation.

Today trains are repaired there – not built. The last coal mine closed in 2015.

▲ *Doncaster still has a market, after more than 750 years.*

▲ *Derelict housing in Doncaster, following the decline of industry.*

Today ... a new lease of life

Today Doncaster has a new lease of life – again because of its location. It is within 4 hours' drive of 91% of the UK's population. So it has become a **logistics hub** for storing and distributing goods.

- It has warehouses for IKEA, Amazon, Next, Lidl, Asda, B & Q and others.
- It shares a new airport with Sheffield, called Doncaster Sheffield airport.
- An inland 'port', the **iPort**, has been built on the site of an old colliery. Amazon and Lidl have warehouses here. The iPort has rail links to sea ports and the Channel Tunnel.

And manufacturing too!

Doncaster still has some manufacturing – but now in new fields. For example, making:

- metal alloys, and components, for planes
- plastic pipes of all kinds, to carry water supply, and underground cables, and for other uses
- machinery for waste recycling
- valves and other engineering products

Doncaster aims to become a centre for **innovation**.

▼ *The end of a shift at the Hatfield colliery. The last working colliery around Doncaster, it finally closed in 2015.*

No. 2 Shaft

Eight hours underground …

▲ A view of Doncaster. Note the railway.

It's logistical ...

▲ Some new warehousing in Doncaster.

Glasgow
Edinburgh
Newcastle
Teesport
Middlesbrough
York   iPort
Leeds   Hull
Humber ports
Manchester
Liverpool   Doncaster
Sheffield
Retford
Derby   Nottingham
Leicester   Peterborough   Haven ports
Birmingham   Norwich
Coventry   Cambridge
Cardiff   Felixstowe
LONDON   Thames ports
Severn ports   Bristol
Portsmouth
Southampton   Channel Tunnel

▲ Doncaster: very well connected!

Key
— motorway
— main trunk road
— railway
● Doncaster, a transport hub
● port
— national borders

So it's alright now?

Not yet. **Unemployment** is higher in the borough of Doncaster, and **wages** are lower, than the UK average. There are still areas of deprivation.

More warehousing is planned, and it will create thousands more jobs. That's good news. But most of the jobs will not need much skill, or pay much.

So the council is working hard to attract more highly-skilled well-paid jobs, and to make sure people are trained to do them. It aims to make Doncaster a great place to work, and live.

Your turn

1   a   Name two activities that flourished in Doncaster as a result of the Industrial Revolution. Give the sector for each.

    b   In which sector are today's warehousing jobs?

2   a   Define *unemployment*. (Glossary?)

    b   What is *an area of deprivation*? (Glossary.)

    c   Explain how the loss of industry can lead to deprivation. Write bullet points? Or draw a spider map, or flow chart?

3   Using the map above to help you, explain why big companies like IKEA are attracted to Doncaster as a warehouse centre. Write at least 8 lines. Don't forget to mention ports!

4   Doncaster aims to attract companies that offer well-paid jobs.

    a   How would well-paid jobs benefit Doncaster? Explain.

    b   Would this help Doncaster to achieve that aim? Explain.
        i   a new university   ii   low-rent offices for IT companies

There are three main reasons why employment patterns change. What are they? Find out below.

What drives change?

As you saw in Unit 3.2, employment patterns change over time. Why? Here are three big drivers of change …

What if …

… farmers did not have machines?

1 Technology

Technology is where we put scientific discoveries and inventions to practical use.

In the primary sector

This big harvester can harvest in a day what it took 10 people more than a week, using scythes, 250 years ago. New inventions in farm machinery have led to falls in the % of workers in agriculture.

In the secondary sector

Modern factories need far fewer workers too, thanks to robots. Here robot arms, controlled by computer programs, are welding cars. So robots contribute to a fall in the % of people in manufacturing.

In the tertiary sector

The computer, the internet, the mobile phone: all three inventions have transformed the tertiary sector since 1970. They allow new ways of working, and new jobs. Up goes the % employed in this sector!

In the quaternary sector

Powerful computers can process vast amounts of data, fast. This has helped the quaternary sector to grow. Tasks that once took years, such as gene sequencing to develop new vaccines, can now be done in days.

2 Globalisation

The second driver of change in employment is **globalisation**.

Globalisation means the increasing movement of goods, services, people, money, and ideas, around the world, making it more connected. Globalisation can have a big impact on employment patterns.

What if …

… robots did everything?

Example: employment in manufacturing

Let's see how globalisation has affected employment in manufacturing, in the UK:

By 1850, the UK led the world in manufacturing. All kinds of goods were made here, and exported. Over 30% of workers were in factories and workshops.

By 1970, many British factories were closing down. Other countries were industrialising, and could make goods more cheaply. British factories could not compete.

Today, many of the goods we buy are imported. Clothing, shoes, textiles, mobiles, computers, fridges, TVs. About 8% of the UK workforce is in manufacturing.

So global trade first boosted our manufacturing, then led to its decline. The decline of traditional manufacturing is called **de-industrialisation**.

But the UK is still world class in some high-value manufacturing, such as:

- **aerospace** (planes, helicopters, fighter jets, satellites, spacecraft)
- **pharmaceuticals** (medical drugs, and vaccines).

3 Governments

A government can also influence employment patterns.

Many people say we need more manufacturing, to balance services. The government can help by offering finance for factories, and grants for research and development. It can work with colleges to make sure young people have the right skills.

In 2019 the government promised to revitalise shipbuilding in the UK. That would be very exciting, and create lots of manufacturing jobs.

▲ The RSS Sir David Attenborough, ready for research in the Antarctic. It was built on Merseyside, and completed in 2020.

Britain once led the world in shipbuilding. The industry declined, but the government has promised to revitalise it.

Your turn

1 Technology is one factor in changing employment patterns.
  a Define *technology*.
  b Give an example of technology that helped to change employment: i 250 years ago ii since 1970

2 a Define *globalisation*.
  b Is this part of globalisation? Give reasons.
    i taking holidays abroad ii using YouTube
    iii buying a T-shirt made in Bangladesh
    iv buying eggs from the farm down the road

3 The % of workers in manufacturing in the UK has fallen, over several decades. Give *two* reasons.

4 A sewing machinist can earn £18 000 a year in the UK – and £1000 a year in Bangladesh. Explain why British chain stores import much of their clothing from Bangladesh.

5 Do you agree with this person? Decide, and give your reasons.

The UK needs more factories!

Most things we buy in the UK are made in other countries. It's one result of globalisation! Find out more here.

Imagine …

Imagine if the only way to reach Britain was by ship, and it took weeks to get news from other countries. That's how life was, 150 years ago.

Today, the UK and other countries are much more connected, thanks to the process of globalisation.

Why has globalisation sped up?

Some say globalisation began with the first trading ships to sail the seas. But the pace has increased rapidly in the last 60 or 70 years. Why? All these have played a part:

▲ *Has globalisation shrunk our world?*

Containerisation Shipping containers hold an enormous amount. They can be stacked up on ships, and moved by rail and truck. They made it easier and cheaper to transport goods.

Information technology (IT) With the internet, and video, and mobiles, you can buy and sell from anywhere, move money in minutes, and get instant feedback.

Air travel Mass air travel began in the 1960s. In 2019 there were over 100 000 flights a day, on average, across the world. Many were international. Planes carry cargo too.

Who pushes globalisation?

- **Businesses** They aim to buy or make goods where it's cheap. They aim to sell goods and services everywhere, to increase profits.

  Some companies have operations (factories, shops, offices) in more than one country. They are called **transnational corporations** or **TNCs**. Many are large and well known. Apple and Samsung are examples.

  The world's TNCs are responsible for *over 50 % of world trade*! They also provide *about 25 % of all employment*!

- **Governments** They favour globalisation, if it helps their economies. For example, the UK is working on trade deals with countries all over the world.

  But note that a government can curb globalisation too. It could refuse work permits to people from overseas. It could impose **tariffs** (taxes) on imported goods, so that they cost more in the shops. Then people buy cheaper local goods instead.

  (Some countries also limit, or ban, access to social media.)

▲ *A street in Shenzhen, China. Spot the TNCs?*

Who wins? Who loses?

Globalisation is complex. There are gains and losses. Let's focus on the UK here.

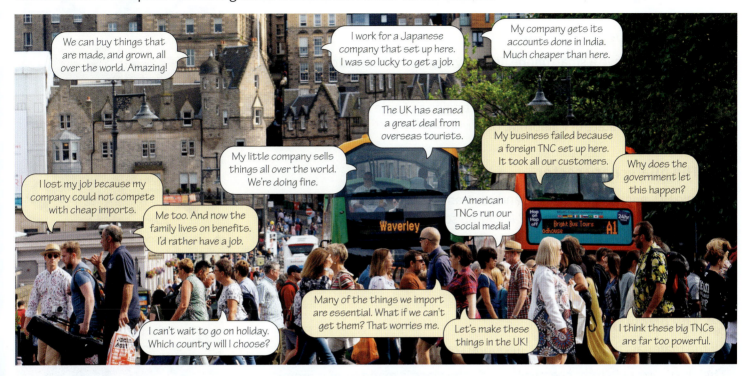

We will look at globalisation and low income countries in Chapter 4.

Will it carry on?

Could the world disconnect again, leaving countries to survive alone? That is hard to imagine. But globalisation *can* be slowed down, or rolled back …

- **by events.** The Covid-19 pandemic that began in December 2019 closed borders, cut air travel, and shut factories. World trade in goods fell. (But trade in some services rose. Netflix gained customers, for example.)

  Climate change may have an impact too. People may decide to travel less, or buy only local goods, instead of goods that travelled thousands of miles.

- **by governments.** They can use tariffs to protect their industries. This may lead to **trade wars**, where other countries do the same, to punish them.

Did you know?
- The USA began a trade war with China in 2018.
- It placed tariffs on many imports from China, including pianos!
- So China did the same on imports from the USA.

Your turn

1. Over 90% of world trade in goods is by sea. Explain why *containerisation* has helped to expand this trade.

2. **a** What do the initials *TNC* stand for?
   b TNCs are all different. What feature do they all share?
   c Give 5 examples of TNCs. (Cars? Entertainment? Food?)

3. From the comments on the photo above, choose:
   a two that show benefits of globalisation
   b two that show harmful impacts of globalisation
   c two that show concerns about TNCs

4. Amazon is a TNC. By the end of 2020, it was operating in 19 countries. Its headquarters are in the USA. It employs over 40 000 people in the UK.
   a Suggest two ways in which Amazon benefits the UK.
   b Amazon has had a negative impact on many businesses in the UK. Explain why.

5. Explain why high tariffs could help to slow globalisation.

6. Look at the first image on page 56 – the coin purse. It aims to sum up globalisation. To what extent does it succeed?

Workers in many countries – and in all four employment sectors – play a part in bringing you a mobile phone. Find out more here.

The story of your mobile

Let's assume you have a smartphone. Its story begins at the headquarters of a company – and moves around the world, taking in every sector.

Let's suppose it's an iPhone. Then the story begins at Apple's headquarters in Cupertino, in the state of California, USA.

1 The quaternary sector

At Apple HQ, engineers, designers and programmers work on features for your phone. What would appeal to you? A fold-up screen? An infra-red camera? Log in by iris? Wave your phone to open your front door?

After many months of top-secret work, the design is perfected.

2 The primary sector

All the materials that make up your phone come from Earth's crust.

The plastic is made from chemicals in oil. The glass is made from minerals extracted from rock.

And then, the metals. Over *sixty* different metals are used in smartphones. Some familiar ones – like gold, silver, copper, tin – and others you may not have heard of – like indium, cerium, and neodymium.

The metals come from all over the world. Many belong to an important group called the **rare earths**, which are mined mainly in China.

3 The secondary sector

The parts for your mobile are made – but *not* by Apple. (Apple does the design and marketing.) They are made in other people's factories, in several countries. Then the phones are assembled, mostly in China.

4 The tertiary sector

Now, the focus is on you. The phones are in Apple shops, and online. The ads are everywhere. People are ready to sell them to you. You are so tempted!

5 The quaternary sector again

You pay for a call plan. You go on the internet, and send texts, videos, audio, and images. You download apps and music. You use GPS. All made possible by clever people in the quaternary sector, in many different companies.

▲ *Earth's resources, processed and neatly packaged, for you to enjoy.*

▲ *A rare earth mine in China. Rare earth metals are used in mobiles, electric cars, and many other products. China controls their export tightly.*

▲ *Getting ready to assemble your phone, in China.*

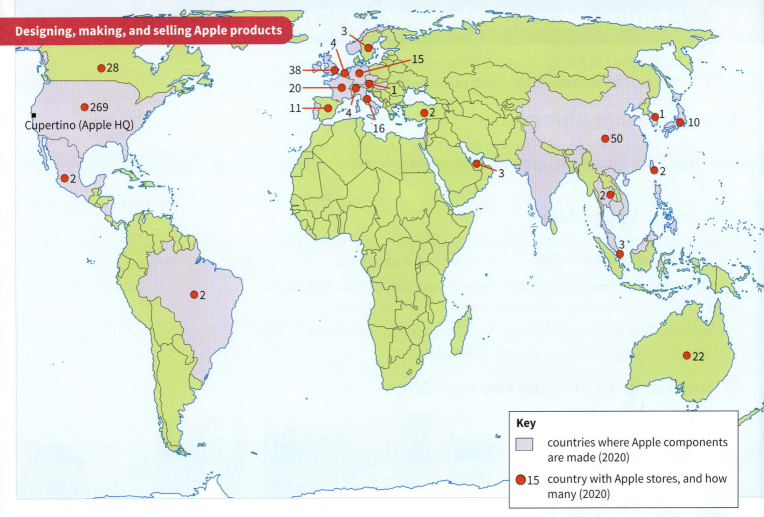

28

269
Cupertino (Apple HQ)

2

2

3
4
38
20
11
4
16

15
1

2

1
50
2
2
3

3

22

Key

	countries where Apple components are made (2020)
●15	country with Apple stores, and how many (2020)

A global company

Apple has its headquarters in the USA, where its global decisions are made. It has stores in twenty countries. It arranges manufacturing in many countries. It sells goods and services everywhere.

Apple's business is a good example of globalisation.

So ... is your mobile linked to globalisation?

Perhaps you don't have a mobile. But if you do, you can be sure that it's made using Earth's resources, from many places. And assembled in countries where wages are lower than in the UK.

▲ *Queuing for a new iPhone in Shanghai. Apple has over 500 stores worldwide.*

Your turn

1 Many different workers play a part in bringing us smartphones.
  a How do workers in the primary sector contribute?
  b In what sector are the workers who ...
    i make the touchscreens? ii sell you the smartphone?

2 a Which country has most Apple stores? Suggest two reasons to explain this.
  b Which European country has most Apple stores?
  c Which inhabited continent has not yet got an Apple store? Suggest a reason.

3 Over 200 suppliers make components for Apple products.
  a Name 6 countries where components are made. (Map!)
  b A large number of Apple's suppliers are in China. What can you deduce about wages in China, compared with wages in the USA?

4 Some American politicians want Apple to have more of its components made in the USA. Why? Suggest a reason.

5 *Apple is a manufacturing company*. True or false? Explain.

A pandemic is a very unwelcome example of globalisation. It affects employment too. Find out more here.

Change, change, change …

As you saw, the balance of jobs *between* employment sectors changes over time. Jobs *within* a sector change too, often because of technology. Change can be speeded up by a shock to the economy – like Covid-19.

The Covid-19 pandemic

Covid-19 was first reported in December 2019, in China. By March 2020 the virus had spread far beyond China, and was declared a **pandemic**.

Covid-19 hit the UK hard. By May 2021, sadly, over 127 000 people had died. Because of lockdowns, hundreds of thousands had lost their jobs.

Covid-19 battered the British economy. It also helped to speed up some changes in the tertiary sector. Let's look at two examples.

Did you know?

- Most pandemics begin with a virus passing from animals to humans.
- Covid-19 has been linked to bats …
- … but it may take years of research to confirm this link.

1 How Covid-19 affected the High Street

The **High Street** means the heart of a town or city, where the main shops are – plus services like banks and restaurants. This photo is from 1958.

Once, shops were family-owned. But the rise of the **chain store** forced most of these to close. Chain stores arrived in High Streets across the UK.

But traffic and parking problems could make it difficult to shop on the High Street. And the shops had to pay a lot for rent and rates.

So **out-of-town shopping centres** began to lure shoppers away. Easy parking *and* all the chain stores, as well as restaurants and cafes.

Then came the next threat to the High Street: **online shopping**. Now you could shop for anything, anywhere, from home. In your bare feet.

In just 9 months in 2019, before Covid-19, over 5800 branches of High Street chain stores closed. They could not compete with online shops.

Some empty shops found new uses: as coffee bars, restaurants, offices, flats, gyms, and leisure centres offering things like climbing walls.

And then, Covid-19. The lockdowns meant High Streets were deserted. People shopped online. Thousands more shops closed down – for good.

So, by causing more shop closures, and boosting online shopping, Covid-19 accelerated the decline in the High Street.

Note that while High Street jobs were being lost, new jobs were being created for online shopping. For example, in marketing, delivery, and warehouse work.

So … what next for the High Street?

Experts predict that the High Street will never return to being rows of shops. It will be a mix of shops, coffee bars, restaurants, apartments, services in health care, fitness, beauty, and other areas – and open green spaces. Watch out for changes on your High Street!

❷ How Covid-19 affected office life

For decades, some companies have let some people work at home part-time. Then, when Covid-19 struck, the government ordered everyone to work from home where possible. Suddenly, office buildings across the UK lay empty.

But with the help of technology – computers, the internet, email, video, mobiles, and services such as Teams and Zoom – office work continued. People proved they could work effectively from home. (Not aways easy!)

So experts predict that in future, working from home – at least part of the week – will be normal. Will this be the case when you start work?

▲ Like office workers, students also had to adapt to working at home.

Your turn

1 **a** What is a *pandemic*? (Glossary?)

  b The Covid-19 pandemic has been closely linked to the globalisation of air travel. Explain why.

2 **a** What does the term *High Street* mean?

  b Which is the nearest High Street to you?

3 Define:

  a chain store (glossary?)

  b out-of-town shopping centre

4 During Covid, more and more people shopped online. Compare online shopping and instore shopping, for:

  a convenience **b** choice

5 Think about a High Street near you (or an imaginary one). What might it be like 15 years from now? Describe it. (You could draw a sketch?)

6 You are an office worker. Suggest:

  a two advantages **b** two disadvantages

  of working from home.

7 During Covid-19, students like you also worked from home. Should working from home become a normal part of the school week? Decide, and give your reasons.

8 *Without technology, the impact of Covid-19 on the economy would have been much greater.* Do you agree? Discuss!

You are likely to work for many years of your life. But doing what? This unit gives you some ideas to think about.

Earning your living

It's fun to imagine what work you might do one day, to earn a living. Many personal factors will affect your choice – such as your interests, skills, qualifications, and where you want to live.

But the other big question is: *What work will there be*? And this is linked to the three big drivers you met already:

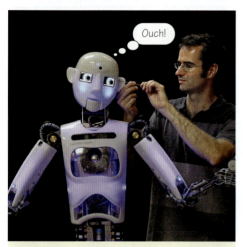

Technology Many of today's jobs did not exist 50 years ago. For example robotics engineer. There will be more new jobs by the time you start work, thanks to technology.

Globalisation For some jobs, you will be able to work for anyone, anywhere in the world, from home. You can use virtual reality, and instant language translation.

The government It can boost science research, through grants. It can encourage manufacturing. It can promote jobs that help to limit climate change.

Any, or all, of these drivers could affect your working life.

Will artificial intelligence help you?

The **Industrial Revolution** began about 250 years ago, and was driven largely by the steam engine. The **Information Technology (IT) Revolution** began about 50 years ago, driven largely by the desktop computer. Both transformed society.

Now the **Artificial Intelligence (AI) Revolution** is underway. Machines are being programmed to do jobs that usually need human intelligence. After a slow start, AI took off around 2000. Will it transform society too?

Robot arms in factories are examples of AI. So are self-drive cars.

You may do a familiar job, but helped by AI. Machines are better than humans at many tasks, for example at picking out tumours on X-rays.

Experts say that AI will take over much of the repetitive work we do – and this may lead to job losses. But AI cannot at this point replace us in jobs that need human contact, and empathy.

In the next level of AI, machines are being programmed to teach themselves, learn from mistakes, plan, and take action. What then, for humans?

What if...

... we had to work only three days a week?

▲ *Alexa: artificial intelligence at home. 'She' combines voice recognition, synthetic speech, and access to a vast amount of data. And she can turn the lights on.*

Will you be part of the gig economy?

The **gig economy** is where you use an **online platform** to find short jobs, rather than permanent ones. Each job is a *gig*.

You access the platform on your mobile or computer. It links people who need a job done with people who can do it, and charges a %.

A platform could specialise in food delivery. So your next gig is to deliver food from restaurant X to family Y, right away. Or it could specialise in graphic design, or plumbing, or legal work, or any other service.

Gig work is growing in the UK. In 2019, 4.7 million people did at least some.

Will you have a green job?

Green jobs are jobs that help to:

- protect the environment, and other species
- limit climate change
- reduce the harmful impacts of climate change

For example, by law, the UK must be **carbon neutral** by 2050, to help limit climate change. We will add *no* carbon dioxide to the atmosphere, on balance. So we must cut back heavily on fossil fuels. Government plans include:

- a big boost for electricity generation from renewable energy sources (wind, solar, wave, and tidal power)
- a ban on the use of all fossil fuels for heating and cooking, in homes
- a ban on new petrol and diesel cars from 2030
- millions of trees to be planted, to soak up carbon dioxide by photosynthesis

Going carbon neutral will take a great deal of work!

Manufacturing wind turbines. Designing houses that need little or no heating. Making films about the environment. Finding bacteria that destroy waste plastic. Becoming a 'green' politician. Those are all green jobs. There are hundreds of green jobs in every sector. Will you have one?

Here comes lunch!

▲ *There goes a gig worker. Gig workers can choose when to work. But they do not know when their next job will come along.*

▲ *By 2050, any power stations in the UK that burn a fossil fuel must use **carbon capture and storage (CCS)**. The carbon dioxide will be captured, and pumped deep underground.*

Your turn

1 It is possible to work from home for a company in another country, even if you can't speak the language. Explain why.

2 What is *artificial intelligence*? Give an example.

3 Robots are being developed to fight fires, using AI. They have sensors for temperature, gas, obstacles, sounds, and movement. What advantages will they have over humans?

4 AI could create problems for humans. In what way?

5 Define the term *gig economy*, and give an example.

6 Think about the gig economy. What are:

   a the benefits   **b** the disadvantages

   for workers? Give as many as you can.

7 What is a *green* job?

8 Explain why this counts as a green job:

   a installing solar panels

   b inventing a way to clear plastic rubbish from the ocean

   c organising protests to ban pesticides that kill bees

9 The UK aims to be carbon neutral by 2050.

   a What does *carbon neutral* mean?

   b Explain how these can help to achieve that aim: farmers   house builders   car factories

10 It is time to design the perfect job for you! What will it be? In which sector? Where? Write at least 10 lines. Add sketches?

3 **Earning a living**

How much have you learned about earning a living? Let's see.

check ✓

1 Page 47 shows a person dressed as Yoda from *Star Wars*.

   a Is this person working for a living? If so, in which sector?

   b From the photo, choose one link between this person and the primary sector, and explain your choice.

   c To which sector do cinema staff belong?

2 Photo **A** shows a typical High Street scene, in the UK.

   a Traditionally, what is the main function of a High Street?

   b Define *chain store,* and find an example in the photo.

   c High Street fashion stores get much of their clothing made in countries like China, Bangladesh, and Vietnam. Why?

   d The international clothing trade is part of the process of *iiablltosoagn*. Unjumble the word.

   e Give two reasons why shops had been closing down in High Streets around the UK, even before Covid-19.

3 Graph **B** is about manufacturing jobs in the UK.

   a About how many workers were in manufacturing:

      i in 1978?   ii in 2020?

   b Describe the trend that the graph shows.

   c Give one reason why the UK lost industry in that period.

   d The UK has some very successful modern industries. Name one area of manufacturing in which the UK excels.

   e About what % of workers were in manufacturing, in 2020?
      2 %   8 %   18 %   28 %   (Page 55!)

4 a What is going on in photo **C**?

   b To which sector does this activity belong?

   c The UK is to be *carbon neutral*, by law, by 2050. Workers at the site in **C** are helping to achieve this aim. Explain how.

   d Give two other examples of *green jobs* that will help us to limit climate change.

5 The gig economy is growing, in the UK.

   a What is the *gig economy*?

   b Give two examples of jobs in the gig economy.

   c A majority of the people who did gig work in 2019 were in the 16 – 34 age group. Suggest a reason for this.

6 Draw a spider map to show the factors that may affect the jobs available in the UK during *your* working life. (Will you include climate change?) Make it look interesting!

7 You are the Minister of Employment. You have in mind the ideal employment structure for the UK.

   a Draw a pie chart to show this structure. (See page 51 for an example.) Mark in the percentages.

   b Explain why you think this structure is ideal.

   c How will you achieve this structure? Give at least two proposals for each sector. (Don't forget quaternary!)

A

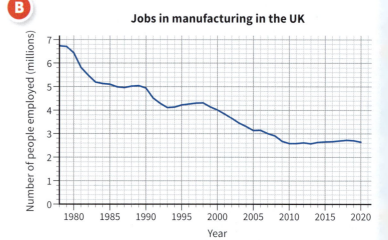

B

Jobs in manufacturing in the UK

(Graph: y-axis: Number of people employed (millions), 0 to 7; x-axis: Year, 1980 to 2020)

C

International development

Have you ever wondered how the world's wealth is shared? Here you can find out.

You did not get to choose …

You did not get to choose the country you live in.
But your country is the main factor in shaping your life. Think about that!
Now imagine that you live in a poor rural village, in a poor country, and …

… this is your family kitchen. No electricity, no running water. One of your chores is to collect firewood.

… this is your nearest water supply, a 15-minute walk away. A full container is heavy. Sometimes the tap runs dry.

… this is the village latrine. It is new. It has no water. Inside is a concrete platform over a hole in the ground.

… this is all your family has for emergencies. You hope nobody gets sick and needs to pay a doctor!

So, much of your life is a struggle. But you are not alone. Many millions of people live in great poverty. And meanwhile, the world's richest people make thousands of dollars a minute. There is great **inequality** in the world.

Poverty, and extreme poverty

Poverty is where you do not have enough resources – money, or possessions – to meet your basic needs. Needs include food, safe water, and shelter.

Extreme poverty is where you have resources worth less than $1.90 a day to live on. (That's about £1.35 – for *everything*.) Barely enough to survive on.

The limit of $1.90 a day is used as a **poverty line** for comparing poverty around the world. Other limits are used too – for example $5.50 a day (about £3.90).

▲ *A poverty line is not about cash to spend. It is about the money value of what you live on. That could include food you grew.*

If the world were a village of 100 people …

Earth has around 8 billion people. Let's imagine it as a village of 100 people. Below are three aspects of this global village in 2019:

E

poverty and inequality

8 villagers lived in extreme poverty, on under $1.90 a day

another 36 were also struggling in poverty, on under $5.50 a day

46 had a better quality of life, with more money to buy what they needed

9 villagers owned a third of the village wealth, between them

and just 1 villager owned half of the village wealth!

F

water supply

33 villagers did not have clean safe water at home – they had to fetch water when they needed it

67 villagers had clean safe water always available at home (piped in, or from their own protected wells)

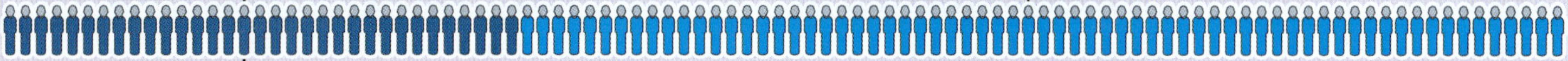

2 of them got unsafe water from rivers and lakes; the rest got water from wells, boreholes, and public standpipes – not always safe; for many, it took at least 30 minutes to fetch water

G

sanitation

8 villagers had no access to a toilet, and used open ground

10 shared latrines with other people

only 45 people – less than half the village – had safe toilets where waste was removed or flushed away, and disposed of safely

8 used holes in the ground, or buckets, or platforms hanging over rivers

29 had their own private latrines

Unsafe water and poor sanitation spread disease. They cause millions of deaths a year – because when poor people fall ill, many cannot afford medical care.

The bad news …

Poverty was widespread in 2019 – but in fact it had been falling for decades.

However, Covid-19 struck at the end of 2019, and brought death, and lockdowns. Jobs were lost. Poverty rose again. For how long? Time will tell.

Climate change is also likely to increase poverty. So we face a huge challenge.

Did you know?

- In 1950, fewer than half of the households in the UK had their own bathrooms.
- Some shared bathrooms and toilets with other households.
- Or you could drag a tin bath into the kitchen, and fill it using buckets and kettles.

Your turn

1 A – D on page 66 show four aspects of the life of a person living in poverty. Study them. Then write four short paragraphs about the same four aspects of *your* life.

2 Look again at A – D. Would it be easy for a person living in this situation to improve his or her life? Give reasons.

3 Define: **a** *inequality* **b** *extreme poverty*
The glossary may help.

4 In **E**, **F** and **G** above, each villager represents 1% of the world's population. In 2019, what % of the world's population lived in poverty – including extreme poverty?

5 Which two statistics in **E**, **F** and **G** did you find the most surprising, or shocking? Explain why.

6 From what you know about *climate change*, suggest two reasons why it may increase poverty in many countries.

Poverty and development are closely linked.
So what does *development* mean? Find out here.

The story so far …

In Unit 4.1, you saw that nearly half the world's people live in poverty.

Every country, including the UK, has some people living in poverty. But a high % of the world's poor live in the poorer countries. Their lives are shaped by the **level of development** of these countries. So let's look at development.

Development has many different aspects

Development is a process of change that improves people's lives. It involves money, of course. But is not just about getting richer. It has many aspects. Compare these:

Did you know?
• The world has more than enough wealth to end poverty.

Why …
… does the world not end poverty?

Aspect	In a highly developed country	In a poorly developed country
poverty	some	a great deal
safe water and sanitation	available to everyone	many have no access to these
education	primary and secondary education for all; a high % go on to college	many children do not even complete primary school; a low % go to secondary school
healthcare	plentiful; easy access to doctors, dentists, hospitals	poor; it may be a very long way to the nearest doctor or hospital – and you may have to pay
roads and other transport links	high quality roads and railways; well connected airports	many roads are just dirt tracks; railways may be rundown; not many flights
employment	low % of workers in farming; high % in services; the key industries produce high value goods	most people live by farming; any industry is likely to produce low value items (like clothes)
% of people living in rural areas	low; most people are urban – they live in towns and cities	high – most people are in farming
fertility rate (average number of children per woman)	low; women tend to have fewer children when they are well educated, and have a career	4 or 5 on average (but the number is falling)
median age of the population	if you line everyone up by age, the person in the middle (the median) is likely to be 40 or over	the median could be as low as 15 (as for Niger in Africa in 2020); a young population

As you can tell from the table, people in highly developed countries are likely to have a higher **quality of life**, with more opportunities and choices.

▼ *Even poor countries have pockets of higher development. This was taken in Kigali, the capital of Rwanda, in Africa.*

▼ *Rwanda is among the world's 20 poorest countries – but developing quite fast, thanks partly to money earned from tourism.*

Trying to develop

There are nearly 200 countries – all at different stages of development.

Most are trying to develop further.

But poorer countries face a huge challenge. It costs a fortune to supply everyone with clean safe water, or a decent education, for example.

So there's a big **development gap** between the richest and poorest countries.

Obstacles to development

Lack of money is a big obstacle. War, and pandemics like Covid-19, can also halt or even reverse development. For example, children could lose years of school.

Climate change will have an impact too.

Different labels

Look at the drawing above. The people represent the world's countries, all at different stages of development. Many different labels get used for different stages. Check out the labels on the drawing. They are grouped by colour.

Did you know?

- The UK is usually in the top 15 most developed countries …
- … but the position of any country can change over time.

Your turn

1 Look at the table on page 68. The second column shows conditions you'd expect in a highly developed country.

  a Pick out the three you think are the most important.

  b i Put your chosen three in what you think is their order of importance, most important first.

     ii Explain why your first choice is the most important.

  c Is the UK highly developed? Decide, and give evidence!

2 Next, look at the third column. If you were in charge of a poorly developed country, which *two* conditions would you want to tackle first? Give reasons for your choice.

3 Look at the drawing above.

  a The slope is quite steep. What is that trying to tell us?

  b The countries are heading towards a better quality of life. Define *quality of life*. (Glossary?)

  c Write your own summary of the message in the drawing.

4 What is an *emerging country*? (Check the labels above.)

5 a What can you say about the UK, from the drawing above?

  b Suggest reasons why its position can change.

6 The photo below was taken in 2019 in Yemen (in the Middle East) after four years of civil war. War can reverse development. Explain why. Would a spider map work?

There are almost 200 countries – all at different stages of development. So how can we compare them? Find out here.

Measuring development

When you visit a country, you can soon get an idea of how developed it is. Just travel around and observe!

But to *measure* how developed it is, you must ask questions like those on the right, and collect **data** to answer them.

Data is collected within most countries every year. It is a big task.

The data is then converted into tables of **development indicators**.

A
1 Do the children get enough to eat, in that country?
2 Does everyone have a clean safe water supply?
3 Can people expect to live long lives?
4 What are the chances of dying when you are only little?
5 Is it easy to get a doctor if you fall ill?
6 Can everyone over 15 read and write?
7 Are people wealthy there – or poor?

What is a development indicator?

A **development indicator** helps to show you how developed a country is.

For example, look at question 3 in drawing **A**. It is answered by a development indicator called **life expectancy**.

Life expectancy means how long people can expect to live for, on average. People in low income countries usually have lower life expectancy. They may have too little to eat, for example, and no access to doctors.

The human development index

A country's wealth, given as **GNI per person (PPP)**, is often used as a measure of development. But on its own, it is a crude indicator. Why? Because the wealth may be shared very unevenly.

So a special index called the **human development index** or **HDI** was created, to compare countries. It looks at three aspects of development:

B Development indicators
- adult literacy rate (%)
- life expectancy (years)
- % of the population with access to clean safe water
- number of doctors per 100 000 people
- under-5 mortality rate (%)
- GNI per person (PPP) ($)
- % of children below age 5 who are underweight

Do you want a game, grandpa?
I'm 92 today.

I went to school for a few months once.
I hope I can go for longer!

Divide this by population to get GNI per person (PPP). Right?

GNI (PPP) $
the country's total *income* for a year, in dollars (PPP)

Life expectancy. This indicates whether people are likely to have a long and healthy life.

Access to education. This is based on how many years of schooling adults have had, and children can expect.

A decent standard of living. This is where GNI per person (PPP) comes in. It is used as the third measure.

The results are combined to give each country a score between 0 and 1. This gives a truer picture of development than GNI per person on its own.

How HDI varies around the world

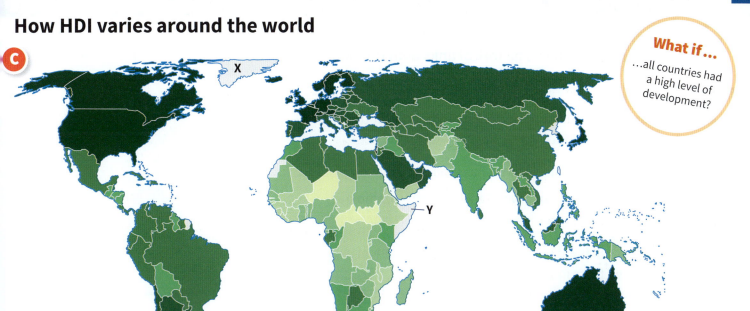

What if...
...all countries had a high level of development?

no data   0   0.2   0.3   0.4   0.5   0.6   0.7   0.8   0.9   1

Map **C** shows HDI values for 2019, before Covid-19 struck. What patterns do you notice?

How does HDI change?

The HDI score rises as a country develops. But it can fall too. For example:

- Covid-19 led to a fall in GNI for most countries. Many older people died. Many children lost months, or even years, of education. The combined effect of all three would reduce a country's HDI, at least a little.

- Climate change is likely to affect the HDI of many countries.

- Civil wars and other conflicts will affect a country's HDI.

Which country am I in?

▲ Her country's HDI gives an indication of the quality of life she can expect – on average. It may rise over her lifetime.

Your turn

1 a Make a table like this, with room for seven questions about a country. Make it full page width.

Question	Indicator	Value in a very poor country (high / low)
1 Do children get enough to eat?		

b Fill in the other questions from drawing **A** on page 70.

c Now fill in the development indicators which answer them. You'll need list **B**, and perhaps the glossary.

d In the third column, write *high* if that indicator is likely to have a high value for a very poor country – or else write *low*.

2 a What does *HDI* stand for?

b A country has an HDI of almost 1. State three things that you can deduce about this country.

3 Use map **C** and pages 140 – 141 to answer this question.

a Which *continent* has most countries with a low HDI?

b Name three countries with an HDI of:
   i under 0.5   **ii** 0.5 – 0.6   **iii** 0.9 – 1.0

4 A country's HDI is 0.375. Deduce three things about the *standard of living* there. (Glossary?)

5 Explain why climate change could lower a country's HDI.

6 Suggest a reason why there is no data for the area on map **C** that is labelled: **a** X   **b** Y

Here we look at the level of development in one country: Malawi, in Africa. It has historical links with the UK.

Meet Malawi

- Malawi is a long thin country in southeast Africa. Look at the map.

- It is about half the size of the UK, in area.

- It has over 19 million people. (The UK has around 67 million.)

- Look at Lake Malawi. It is shared with Tanzania and Mozambique.

A little history

Malawi was once part of the Maravi Empire (around 1500 – 1890). The first British person arrived there in 1859. He was Dr Livingstone, a Scottish doctor and missionary. He was followed by other missionaries, and traders.

In 1891 Britain took control of Malawi. British planters set up plantations there, to grow tobacco, cotton, and other crops, for export. But in 1964, after decades of struggle, the country gained independence.

How is Malawi doing?

- Malawi is one of the world's poorest countries. Around 51 % of its people live in poverty, with 20 % in extreme poverty.

- Nearly 80 % of its workers earn a living by farming.

- Tobacco is its top export. It also exports sugar, tea, and cotton.

- It has few mineral resources (but there may be oil under Lake Malawi).

- Compare this data for Malawi and the UK, for a recent year:

Values for ...	Malawi	UK
GNI per person (PPP)	$1035	$46 070
HDI	0.483	0.932
Average years of education	4.7	13.2
Life expectancy (years)	64	81
Fertility rate	4.1	1.7
Number of doctors per 100 000 people	4	281
Median age of the population	18	41

Now read about Sephora, on the next page. Then try *Your turn*.

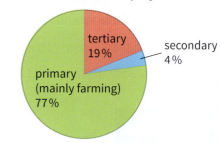

Malawi: structure of employment

▲ *Employment structure for Malawi in a recent year. Compare with the UK (page 51).*

Your turn

1 a Where is Malawi? Give details. (Page 141?)
   b For how many years did Britain control Malawi? Calculate!

2 Look at the data in the table above. How many times richer is a person in the UK, on average, than a person in Malawi?

3 How many years longer will a person live, on average, in the UK than in Malawi? Suggest two reasons for this difference.

4 Compare employment structures for Malawi and the UK.

5 Explain how this could have held back Malawi's development.
   a It is a *landlocked* country. (Glossary?)
   b On average, people spend 4.7 years in education.
   c The population grew from 11 million in the year 2000 to over 19 million in 2021.

6 Page 73 tells about Sephora's school day. Describe five ways in which *your* school day is different from Sephora's.

Sephora's day

Sephora lives near the northern end of Lake Malawi. She's nine.

Off to school

It's 6.15 am. Sephora is on her way to school, in her worn dress and bare feet. She sings as she walks along the dirt track. The sun is already bright, and the mountains gleam in the distance.

School starts at 7 am. After outdoor assembly she heads for her classroom, with 97 other children. There are no desks, so she'll sit on the floor. She has a pencil, but her exercise book is full. Her mum can't afford another one yet. So she won't be able to write anything today.

The teacher works hard, and does her best. But it's hard to teach 98 children – or even remember their names!

After school

School finishes at 1 pm. Sephora runs home to help on the farm. She will weed the vegetable patch, look after the hens, and get water from the well. And later, help her mum make maize porridge for the evening meal.

Then it gets dark ...

It gets dark around 6 pm. There's a kerosene lamp and a torch. But these don't give much light. So by 7 pm, Sephora is lying on her mat on the mud floor with her little sister, ready for sleep. Her two brothers are in the corner, on their mat. Outside, the dogs bark and the frogs croak.

What will her future be?

Sephora thinks over what her dad said. 'This will be your last year at school, Sephora. You'll stay at home, and the boys will start school.'

She's sad. She longs to complete primary school, and go to secondary school. She knows she learns fast. But now she won't have the chance.

So what will become of her? She'll work on the farm. Then get married at 18, and have her own children. A girl in the village got married last week, at 14! It's illegal – but it happens. She hopes her dad would not allow it.

Hello.

Where are you going?

▲ *In Sephora's village.*

▲ *In Malawi's primary schools, average class size is over 80. (Not enough teachers!) The students could excel, like students in the UK – but they don't have the same chance.*

▼ *Elephants, hippos, rhinos, monkeys, zebra: Malawi has them all.*

▼ *Lake Malawi: many people earn a living from fishing.*

Why are some countries much more developed than others? This unit gives some reasons.

The development gap

As you saw, some countries are much wealthier and more developed than others. Why? In this unit and the next, we look at some reasons.

What if …

… the UK was one of the world's poorest countries?

Did you know?

• China is predicted to replace the USA as the world's top economy by 2032.

1 Historical reasons

500 years ago, Asia was the richest continent – about twice as wealthy as the rest of the world combined, and much wealthier than Europe.

But things change. And by 1750, the **Industrial Revolution** had begun in Britain. It spread, and Europe leaped ahead in wealth and development.

Europeans had already settled in North America. So industries began to develop there too – and North America began to grow wealthy.

Meanwhile, Europeans had been exploring Africa, South America, and Asia. They had found lands rich in natural resources. Trading soon followed.

The Europeans traded for things like gold, ivory, tobacco, spices – and in some places, slaves. But as time went by they grew greedier …

… and took over places as **colonies**, to gain control. The materials and enslaved people they shipped made many Europeans very rich!

Eventually, the colonies won their independence. But most were left with few roads, schools, hospitals, or skills … and much unrest.

Overall, several European countries – including Britain – grew richer, and more developed, by exploiting their colonies. But they did little to develop them. Many ex-colonies, such as Malawi, are still very poor today. Some are unstable. (But some, like Singapore, are doing fine.)

2 Geographical reasons

A country's location, and climate, and natural resources, can play a big part in helping it to develop.

In a hot dry landlocked country, with poor soil and few other natural resources, development may be very difficult.

But some countries have rich soil, and a climate good for farming. And natural resources, such as oil and ores, that other countries want to buy.

Some benefit from their location. The tiny island country of Singapore is an example. It sits in one of world's busiest shipping routes.

3 Health and education

A well-educated, skilled, and healthy workforce helps a country to develop. People can come up with bright ideas, and work together to tackle problems. But poor countries are at a disadvantage.

Poor countries have lots of bright young people. But many don't get a good education. (There aren't enough trained teachers, for one thing.)

Diseases such as malaria, TB and AIDS are common in many poor countries. If you are unwell, and undernourished, you can't work.

If you are poor, most of your energy will go to finding food, and water, and firewood. You won't be able to think about much else.

Your turn

1 A – D are about different countries. Explain how each could have held back development.

   A It has a very poor education system.

   B It was a colony for nearly 150 years.

   C Malaria is very common there, and treating it costs a lot.

   D It is hemmed in by mountains, and hard to reach.

2 Of the four situations described in 1, which do you think could be improved, to help that country develop? Explain your choice.

3 The UK is among the world's most developed countries. Using ideas you met in this unit, give five reasons to explain why. (Don't forget geographical reasons.)

Here we give more reasons to explain why poor countries find it hard to escape from poverty.

Keeping the gap wide

In Unit 4.5, you met some reasons for the development gap between countries. Now let's look at other things that help to keep poor countries poor.

Why …
… do we still go to war?

Did you know?
- It's estimated that about $90 billion leaves Africa every year through corruption.

4 Conflict and corruption

Hi ho, hi ho …

… it's off to work we go.

A country has a better chance of developing if it is at peace, with a stable government, and a strong and fair legal system.

Those new tanks must have cost a fortune.

Enough to build a few hospitals.

But many poor countries do not have stable governments. Many are deep in **conflict**, with a big waste of life – and money.

Next stop, my Swiss bank account.

In many countries, **corruption** adds to the problem. Leaders and officials take bribes, and steal money that should be spent on development.

There is some corruption in every country. But it is widespread in some of the poorer countries, and has a major impact on development.

5 Relying on a few exports

Usually, countries earn money by selling things to other countries.

Good crop.

I hope it gets a good price.

Many poor countries rely heavily on selling just one or two **cash crops** to other countries, to earn money. But that's risky …

Look. Only £20 a tonne.

No, over here! £10 a tonne!

… because the amount they earn for a crop can tumble. For example, if other countries decide to grow it too. Or if demand for it falls.

SMOKING KILLS

For example, tobacco is Malawi's main cash crop. But more people are giving up smoking, so the demand for tobacco is in decline.

The same is true for other **commodities**. For example, the demand for oil will fall as we tackle climate change. So its price will fall. Oil producing countries will suffer. It is risky to rely heavily on exporting a commodity.

6 Lack of industry

You can usually earn more by selling factory goods than crops and raw materials. For example, suppose your country grows cocoa.

You can sell your cocoa on the world market. Chocolate companies in richer countries will buy it to make chocolate.

The chocolate companies like to buy cocoa cheaply, if they can. But they charge quite a lot for chocolate – and can charge more every year.

So their profits rise, while your earnings may fall. How nice it would be to make your own chocolate, and export it!

But it costs a lot to set up factories. Poor countries may not have the money or expertise to get started. Even if they do, the electricity supply may be unstable. Or poor roads may make the transport of goods difficult. (Colonisers did not invest much in their colonies!) So it's hard to run a factory efficiently.

▲ A cocoa farmer in Ghana. Those yellow pods contain cocoa beans, which are used to make chocolate.

▲ When you buy chocolate, the cocoa farmer will have received only a small fraction of what you pay.

Your turn

1 Define these terms. (Glossary?)
   a *corruption*   b *commodities*   c *adding value*

2 A – C are about different countries. Explain how each could have held back development.
   A The leader stole £millions meant for new roads.
   B It depends almost completely on exporting cotton.
   C It has few factories, and imports almost everything.

3 Setting up chocolate factories could help a cocoa-growing country to develop. Explain why. You could draw a flow chart.

4 Poor infrastructure helps to keep a country in poverty.
   a What is *infrastructure*? (Glossary.)
   b Explain how this could hold back development:
      i poor roads
      ii an unstable electricity supply

4.7 Globalisation, development, and TNCs

Big companies called TNCs can move in, and speed up a poor country's development. Or can they? Find out here …

Globalisation and development

Globalisation is the process of creating a more connected world, through flows of trade, money, people, and knowledge. (Check back to Unit 3.4.)

Globalisation can help a country to develop.

For example, the Industrial Revolution made Britain the world's top manufacturing nation. It sold goods all over the world. The money it earned helped it to develop.

Today, China is the world's top manufacturing nation. It sells goods all over the world. The money it earns is helping it to develop – fast! Over 850 million Chinese people have been lifted out of poverty since 1980.

The role of TNCs

Globalisation is largely driven by **transational corporations**, or **TNCs**. These are companies with branches and operations in more than one country. (See page 56.) They are responsible for *over half of all world trade*. So they can have a huge impact on development.

Many TNCs from wealthy countries operate in low income countries. Why? Because they usually aim to make as much profit as possible. They want to:

* obtain materials as cheaply as possible (for example, metal ores)
* benefit from workers with low wages
* be in, or close to, large populations to sell things to.

So a foreign TNC might open mines in a developing country, to exploit its minerals. Another TNC might set up a factory to make clothing cheaply.

The pros and cons

When TNCs operate in a country, they pay taxes to the government, and wages to their workers. The money can help the country to develop. That sounds perfect? But the story is far more complex. It is not all good news. Study the next page. Then try *Your turn*.

▲ *China: rapid development thanks to TNCs, industry, and globalisation. Development took off in 1980 when it invited foreign TNCs to set up factories there.*

▲ *Bangladesh: developing thanks to globalisation and garments! Clothing makes up 80% of its exports. Most fashion TNCs get clothing made in factories there.*

Your turn

1 British Petroleum (BP) is a TNC. It operates on every continent.
  a What is a *TNC*?
  b What is the main reason a TNC sets up in a new country?
  c Give two other examples of TNCs.

2 Look at the points in *favour* of TNCs on page 79. Which two do you think would be the most important, if you were:
  a a government minister?   b a mum with a young family?
  For each, explain your choice.

3 Now repeat question **2** for the points *against* TNCs.

4 Suppose you live in a low income country. What do you think your *overall* attitude to TNCs will be, and why, if you are:
  a a doctor?   b trying to start your own factory?

5 Do you agree with this person? Decide. Then write a response to justify your decision. Try for at least half a page!

Developing countries should keep TNCs out!

In favour of TNCs in developing countries

Against TNCs in developing countries

The Covid-19 pandemic affected the whole world. But richer countries were first in the vaccine queue. Find out more here.

The virus that shook the world

On 31 December 2019, China reported an outbreak of disease caused by an unknown virus. It was named Covid-19, after the year it was found.

The virus spread fast from country to country, carried by infected travellers. In the early months, it spread fastest to the richer countries, that were highly connected by air travel.

On 11 March 2020, Covid-19 was declared a **pandemic**.

Dealing with the pandemic

Experts had warned the world for years that there would be pandemics. In 2005, 196 countries had signed an agreement to prepare for them.

But when Covid-19 struck, few countries were ready. Many had no clear plan. Politicans argued for months about what to do:

▲ The fastest way to travel. Was Covid-19 on board? Did it spread in the plane? And when the passengers landed …?

As the weeks went on, scientists told us more about how the virus spreads – in droplets carried in breath and sneezes. Most governments ordered mask wearing, and lockdowns. Travel between some countries was banned.

The vaccine race

It was clear that the only way to end the pandemic was to find a vaccine. But vaccines usually took 10 – 15 years to develop. Too long!

Thanks to brilliant work by scientists, the first vaccines were ready in under a year. By May 2021 there were 13 in use – developed in different countries, including the UK, USA, China, India, and Russia. More were on the way.

How were they developed so fast? Because …

- teams around the world helped each other, instead of competing.

- advanced tools and techniques in the quaternary sector were a big help.

- money poured in from high income countries, to fund the research. They knew a vaccine would help their economies to recover.

- many countries agreed to take part in the vaccine trials.

So the vaccines were a global effort. Would they be shared fairly? Read on…

▲ A Chinese vaccine being trialled in Brazil. Before vaccines are approved, they must go through trials using thousands of volunteers, to make sure they work, and are safe.

Did you know?

- It has taken over 40 years to find a vaccine for malaria.
- Two were in trials in Africa in 2021.

The early scramble for vaccines

The richest countries were first in the vaccine queue.

- They ordered vaccines in advance from the drug companies, even before approval. They competed to get them.

- They ordered more than they needed, in case some vaccines failed, or did not work for long. The UK ordered about three times what it needed.

- By December 2020, rich countries with 14 % of the world's population had pre-booked over half of the supplies for 2021.

Poor countries could not compete. But some got free vaccines from China and other countries, who wanted closer ties. This is **vaccine diplomacy**!

▲ *Packing Covid-19 vaccines at the Serum Institute in India. It is the world's largest producer of vaccines. It makes them on behalf of drug companies – and also develops its own.*

COVAX: a way to share fairly

Until every country had vaccinated its people, no country would be safe. So a scheme called **COVAX** was set up, to share vaccines fairly. Like this:

| Richer countries pay money into COVAX. | → | COVAX uses the money to help fund the development and production of vaccines. | → | COVAX also buys the vaccines; it can get a low price because it can buy billions of doses. | → | COVAX shares the vaccines equally among all countries; the poorest 92 countries get them free. |

COVAX hoped to give the 92 poorest countries enough vaccines for 20 % of their populations, in 2021. Good – but not enough. Delays mean more time for the virus to spread, and **mutate**. Then those vaccines might not work.

Given the slow delivery to poorer countries, it was likely to take over two years to vaccinate everyone.

Suppose richer countries had not rushed to buy vaccines, but gave more money to COVAX instead. Would the world be protected sooner?

What about the future?

- Covid-19 is not going away. There will still be outbreaks, and new mutants. We will live with it, protected by vaccines.

- There will be more pandemics. That is certain. But countries may have learned from Covid-19, and be better prepared.

- Scientists have learned a great deal about new ways to make vaccines.

- COVAX may become the standard for future vaccines. A good idea?

▲ *The first COVAX delivery! Vaccines arrived in Ghana (Africa), by air, on February 24, 2021. The first priority: to vaccinate healthcare workers.*

Your turn

1 Air travel helped Covid-19 to become a *pandemic*. Explain why.

2 When Covid-19 struck, many countries were reluctant to:
  a close their borders
  b lock down businesses
  c close schools
  For each, explain why.

3 Suggest three reasons why poorer countries find it harder to cope with a pandemic than richer countries do.

4 From what you know already, suggest two reasons why:
  a most vaccines were developed in richer countries
  b vaccinating everyone in poor countries is a big challenge

5 If the world's low income countries were not vaccinated …
  a their poverty levels would increase. Explain why.
  b the economies of the richer countries could suffer. Why?

6 Is COVAX a good idea? Share your thoughts!

If you lived in poverty, with little hope, would you try to escape to seek a better life? Many millions do. Read on …

Seydou's story

It's four months since I said goodbye to mum and dad in Mali. I miss them so much. But here I am, in a migrant centre in Lampedusa.

I thought it was a great idea to come to Europe. I'm 17, and I can work really hard. I need to earn money, to help my family like I promised. But I didn't know the journey would be so tough, or that I'd see such terrible things.

My land journey

I went to Tamanrasset first, in Algeria. I got work with builders there, to earn money for the next stage. Then to Sebha, and more work. But I was beaten up in Sebha, and some of my money was stolen. So when I got to Zuwara, I had to find work again. The people smuggler charged a lot. I hid in the dark for two days, in a shed with over 100 other people, till he was ready to pick us up.

My sea journey

The boat was packed. People sea-sick, people praying, babies crying. Water started coming in. We kept bailing it out. When we got near Lampedusa we saw the rescue boat. People stood up on that side, and waved and shouted. The boat nearly capsized. People fell out – maybe eight or nine. Two were children. None could swim. We lost them.

Where next?

I'm lucky I got this far. But I am scared. Will they let me stay in Europe, or send me back to Mali? Mum and dad must be worried sick. And they need me to send money. They have hardly enough to eat.

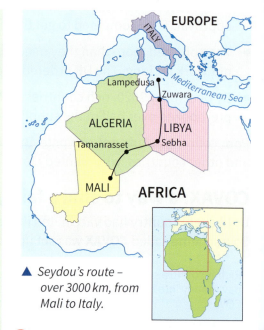

▲ Seydou's route – over 3000 km, from Mali to Italy.

▲ He took what transport he could afford.

◀ Saved by a rescue boat. These migrants will be taken to the Italian island of Lampedusa.

▼ A protest in London.

One Migrant Drowning is a tragedy Thousands Drowning is Foreign Policy

UK RESUME SEARCH · RESCUE FOR MIGRANTS

Seydou was not alone

- Around 3 million migrants crossed the Mediterranean from North Africa to Europe in the years 2015 – 2020. At least 20 000 died on the journey, most by drowning.

- In the first 4 months of 2021, 17 360 crossed alive, and 599 drowned.

Why do they do it?

- Like Seydou, many people make the journey across the Mediterranean to escape poverty, and help their families.

- Many others are fleeing from conflict. Many travel via North Africa to escape conflict in Syria and other countries.

- All know they risk their lives – but they feel they have no option.

- All face an uncertain future. They have no visas or other permits. Will they be allowed to stay in Europe? Or forced to return home – or hide?

It's not just the Mediterranean

- All over the world, *as you read this*, thousands of people are risking everything, without permits, for a better life in another country.

- For example, many are trying to cross the border from Mexico to the USA. (Look at the map on page 140.) Some will perish in the deserts of Arizona and Texas. Some will be arrested by border patrols.

- In Asia too, people try to reach wealthier Asian countries – or Australia.

The big dilemma

The migrants' dream of a better life is causing a dilemma.

- Some people say the migrants must be sent home again – unless they came from war zones, and were truly in danger.

- Others say we must show compassion to all migrants. We don't choose where we are born. Rich countries can afford to take migrants in.

- In fact the richest countries help less than others. Many African countries have huge numbers of refugees. Turkey is a middle income country in Asia. In 2020 it had the world's largest refugee population – about 4 million.

One solution is to help poorer countries out of poverty. Then people would not need to escape. But how? Find out in the next unit.

▲ *The USA built a wall along part of its border with Mexico, to stop migrants entering illegally. Now they try more dangerous routes.*

▲ *Volunteers leave food and water on the USA side of the border, to help migrants crossing from Mexico. (Hundreds die in the desert of thirst and heat stroke every year.)*

▲ *A protest in the USA about policy towards migrants entering illegally from Mexico.*

Your turn

1 a Why did Seydou make his journey?
   b Seydou's journey took several months. Why?

2 a Seydou's boat sailed from *Zuwara* to *Lampedusa*. Which countries are those places in?
   b Both are key places on migrant routes from Africa. See if you can explain why. Page 141 may help.

3 Why do people smugglers do what they do?

4 a Explain these terms. (The glossary may help.)
      i   *migrant*   ii *economic migrant*
      iii *refugee*   iv *entering a country without a permit*
   b To which groups in **a** does Seydou belong?

5 a Look at photo **F**. Work out what the message on the placard really means. Then rewrite it in your own words.
   b Now do the same for photo **C**. It's more of a challenge!

This unit is about ways to end world poverty.

The big challenge

As you saw in Unit 4.1, billions of people live in poverty. Of these, hundreds of millions live in *extreme* poverty, surviving from day to day on almost nothing.

Does it have to be this way? No! So what can be done?

1 Poorer countries can help themselves

A Poorer countries can try to exploit more of their natural resources, to earn more. For example, develop tourism, or grow new crops.

B They can root out corruption. And build more schools, and hospitals, and roads, and other infrastructure. They can educate and train people.

C They can develop manufacturing. Factories can provide steady work, and a steady wage. The goods can be exported.

Development led by a government is called **top-down development**. Development schemes cost a lot. Governments can involve foreign TNCs in some projects. But they also need help from richer countries.

2 Richer countries can help them

Richer countries can help poorer countries by giving them **aid**.

D Richer countries give aid in the form of grants, or cheap loans, or help and expertise for big projects like roads and railways.

E For example, China – not yet a high income country – gives a great deal of aid. China has built roads, hospitals, and much else in Africa.

F But poorer countries say trade can help even more than aid: we should buy the crops and other goods they produce, at a fair price.

In 1970, the rich countries agreed to give 0.7 % of the wealth they produce each year, as aid to poorer countries. But most give less than that.

3 Help from NGOs

£100 is my target.

Non-governmental organisations or NGOs help development through small local projects. They raise money from people like you!

For example, they may provide the materials and expertise for villagers to dig a well, so that they have clean safe water to drink.

Or give people small loans, to help them set up in business. That's called **microfinancing**. Abina got a loan to buy a sewing machine.

This kind of development, to help a local community, is called **bottom-up development**. Small projects can transform lives. Every little helps!

4 Help from technology

Mobile phones are transforming lives in low income countries. For example, if you are a farmer, you can quickly get advice about selling crops. You don't need the electricity grid to charge your phone. Use a solar-powered charger!

You can transfer money by mobile too. You go to an agent and pay money into your mobile account. Then text a code to the person you want to pay. He or she takes the code to another agent, and gets cash in return. This service, launched first in Kenya, is helping millions of Africans.

A little more about manufacturing

You know already that manufacturing can play a big part in helping a poor country to develop. Foreign TNCs often set up the factories.

Low income countries usually start with factories for low value, high volume goods that need a lot of workers. For example, clothing and shoes. As a country develops, manufacturing usually shifts to higher value goods.

Got it!

▲ At the agent's office, with a code on his mobile to pick up cash. The mobile money service has spread fast across developing countries. No need for a bank account!

Your turn

1 Look at the text below photo **A**. You could summarise it as: *Exploit natural resources*.
Summarise the text under each photo **B – I** in the same way. Not more than one line for each!

2 Choose any *two* photos from these: **B, C, F, I**.
For each, explain how the action(s) you summarised in **1** can help to lift people out of poverty. Would flow charts help?

3 Define, and give one example of:
  a **i** *top-down development*   **ii** *bottom-up development*
  b Which type of development in **a** might *you* help with? Explain your answer, with the help of an example or two.

4 She has a question for you.
Write a thoughtful reply.

Why bother helping poorer countries?

4 International development

check ✓

How much have you learned about international development? Let's see.

1 Photo **A** was taken in Liberia. It is a small country on the coast of West Africa. The woman is called Ellen.

  a Find Liberia on the map on page 141, and name its capital city.

  b The photo shows Ellen's home. Name three *utilities* in your home, that Ellen's home is not likely to have. (Glossary?)

  c The population of Liberia is around 5 million. Roughly half live in extreme poverty. Define *extreme poverty*.

  d By 2020, only 12 % of Liberians had access to electricity. Give one way in which this could hold back development.

  e An NGO will help Ellen's village to install solar panels. This is called _____ *development*. What are the missing words?

  f Now turn to the map on page 71 and find Liberia. Which HDI range is it in?

2 Drawing **B** represents one of the development indicators you met in this chapter. Identify the indicator.

3 **C** compares indicators for Armenia (in Europe) and Equatorial Guinea (in Africa). Their GNI per person is almost the same.

  a What does *HDI* stand for?

  b Draw a sketch or spider map to show which aspects of development are combined, to calculate HDI.

  c Which of the two countries in **C** has a higher HDI?

  d Which country gives children a better start in life?

  e There is extreme inequality in one of the two countries. The people in power have vast wealth, but much of the population lives in poverty.
Which country? Explain how you decided.

  f Which indicator gives a truer picture of development?

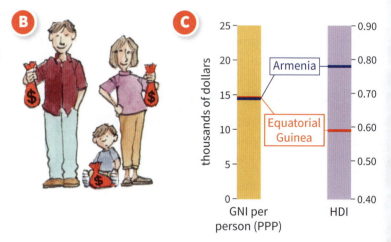

4 Graph **D** compares HDI for four countries from 1990 to 2019.

  a What is the overall trend in HDI for all four countries?

  b Is this trend a good thing, or a bad thing? Explain.

  c China has a huge population (over 1.4 billion). Between 1990 and 2019 it lifted hundreds of millions of its people out of poverty. What was the main reason it could do this?

  d China and India are described as *emerging economies*. What does that tell you? (Glossary.)

  e Malawi has a very small population compared to China. So you might expect it to develop faster! But no.
Give: **i** one geographical factor **ii** one economic factor that has helped to hold back development in Malawi.

  f In 2002 there was a famine in Malawi, because the rains failed. *How* and *why* did this affect Malawi's HDI?

  g Comment on the UK's HDI.

5 In 2018, a Chinese TNC announced plans to set up factories in Malawi, making cotton clothing and towels.

  a What is a *TNC*?

  b Suggest *four* reasons why it chose Malawi. (Page 72?)

  c The president of Malawi was thrilled at this decision. Why?

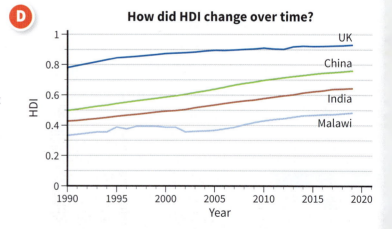

How did HDI change over time?

6 Now turn back to the girl on the tightrope on page 65.

For a country, development is like walking a tightrope, in some ways. What kinds of things can knock the process off track?

7 *A low income country needs other countries to help it develop. To what extent do you agree with that statement? Justify your answer. Write at least half a page. Don't forget trade!*

5 Our restless planet

You know quite a lot about the outside of Earth, where you live. But what's it like inside? Find out in this unit.

Earth's three layers

Earth is almost a sphere. (It is a bit fatter at the Equator.) Its radius is about 6400 km. That's about twelve times the distance from London to Edinburgh.

And it is not all solid! It is made up of three layers:

① **The crust**
You live on the crust. It is Earth's hard skin of rock. At the UK, it is up to 35 km thick.

② **The mantle**
It is about 2900 km thick, and made of solid rock. The rock in the **upper mantle** is hard and rigid. But the rock in the **lower mantle** is not rigid. It can creep or flow very, very slowly.

③ **The core**
It is made of metal – mainly iron, mixed with nickel. The **inner core** is solid. The **outer core** is liquid.

the crust

the mantle

the core

1260 km

2220 km

2900 km

The crust is 7 – 70 km thick.

The rock is solid and rigid in the upper mantle …

… and solid but not rigid in the lower mantle.

The outer core is liquid (iron + nickel).

The inner core is solid (iron + nickel).

How did the layers form?

Some time after Earth formed, it got so hot that everything inside it melted. The heavier substances in the liquid sank, and the lighter ones rose, making layers. As Earth cooled, the layers remained.

Still hot …

It's still very hot inside Earth. It gets hotter as you go down through it. The temperature at the centre is estimated to be around 6000 °C. That's more than enough to melt iron and nickel. But the high pressure stops them melting.

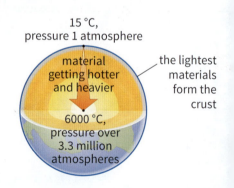

15 °C, pressure 1 atmosphere

material getting hotter and heavier

the lightest materials form the crust

6000 °C, pressure over 3.3 million atmospheres

▼ *Boiling rock reaches Earth's surface in many places. This is the Erta Ale lava lake in Ethiopia.*

▼ *One way to find out about Earth's interior is to drill holes. Under this tower on the Kola Peninsula in Russia is the deepest hole drilled so far – 12.3 km. Imagine it!*

More about the crust and mantle

This drawing shows part of the crust and mantle (not to scale).

The crust under the oceans is called the **oceanic crust**.
The crust that forms the land is called the **continental crust**.
The rock in the oceanic crust is heavier.

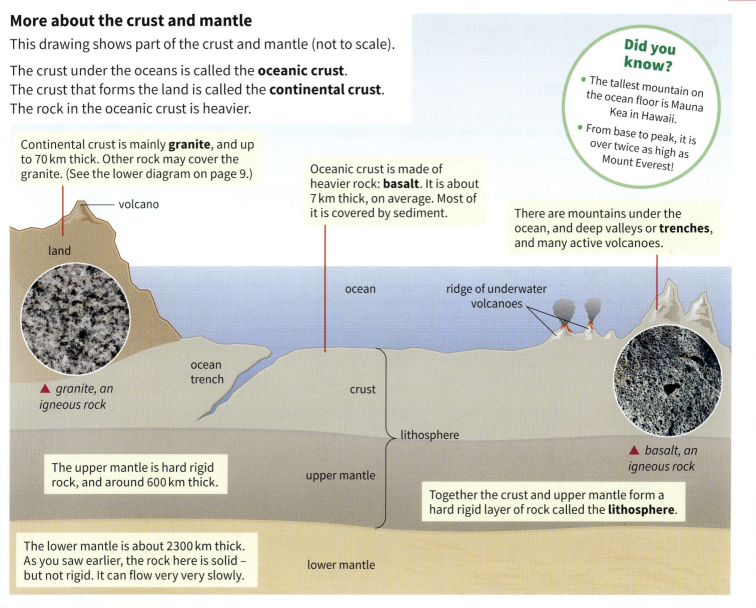

Continental crust is mainly **granite**, and up to 70 km thick. Other rock may cover the granite. (See the lower diagram on page 9.)

volcano

land

▲ *granite, an igneous rock*

ocean trench

The upper mantle is hard rigid rock, and around 600 km thick.

The lower mantle is about 2300 km thick. As you saw earlier, the rock here is solid – but not rigid. It can flow very very slowly.

Oceanic crust is made of heavier rock: **basalt**. It is about 7 km thick, on average. Most of it is covered by sediment.

ocean

crust

upper mantle

lower mantle

ridge of underwater volcanoes

lithosphere

There are mountains under the ocean, and deep valleys or **trenches**, and many active volcanoes.

▲ *basalt, an igneous rock*

Together the crust and upper mantle form a hard rigid layer of rock called the **lithosphere**.

Did you know?
- The tallest mountain on the ocean floor is Mauna Kea in Hawaii.
- From base to peak, it is over twice as high as Mount Everest!

The lithosphere floats on top of the lower mantle. And as you'll soon see, it's on the move!

Your turn

1 Make a table like this, and fill it in for Earth's layers.

Layer	Made of ...	Solid or liquid?	How thick?
crust	rock		
mantle – upper			
mantle – lower			
core – outer			
core – inner			

2 **a** The radius of Earth is about _____ km?

   b If you could cycle non-stop to the centre of Earth, about how many days would it take you, at 20 km an hour?

3 Explain what these are, and how they differ:

   a continental crust   **b** oceanic crust

4 Describe Earth's lithosphere. Include the word *rigid*!

5 What is unusual about the solid rock in the lower mantle?

6 Copy this summary diagram, and complete the labels.

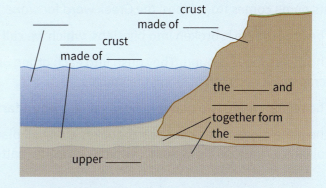

_____ crust made of _____

_____ crust made of _____

the _____ and _____ _____ together form the _____

upper _____

📍 Earth's rigid outer part – the lithosphere – is cracked into big slabs.
These are linked to earthquakes and volcanoes! Find out how.

First, a puzzling pattern

An **earthquake** is caused by rock suddenly shifting.
A **volcano** forms when liquid rock bursts out through Earth's hard surface.

Map **A** shows Earth's main earthquake and volcano sites. What do you notice?

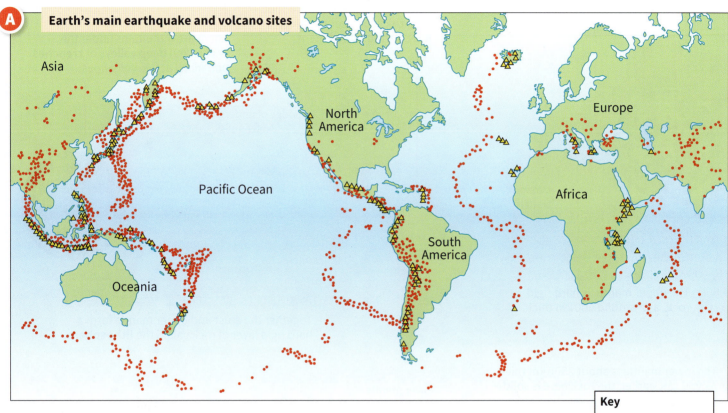

A Earth's main earthquake and volcano sites

Asia · North America · Pacific Ocean · Oceania · South America · Europe · Africa

Key
- • earthquake site
- △ volcano

As you can see, the sites of earthquakes and volcanoes:

- tend to lie along lines – they are not just random

- are often found together

- occur in the ocean as well as on land.

It took years for scientists to explain this puzzling pattern.

Explaining the pattern

- As you saw earlier, the **lithosphere** is the hard rigid outer part of Earth. (Its name comes from *lithos*, the Greek word for *stone*.)

- The lithosphere is broken into big slabs, which we call **plates**.

- The plates move around. They push into each other, or pull apart, or scrape past each other, along their edges.

- The region where two plates meet is called a **boundary** or **margin**.

- The plate movements cause earthquakes along all plate margins, and volcanic eruptions along some. And this gives the pattern in **A**.

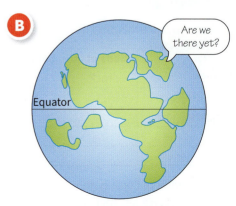

▲ *Earth a billion years ago? Its plates have been moving, joining, and breaking up for billions of years. So the continents and oceans have changed shape too.*

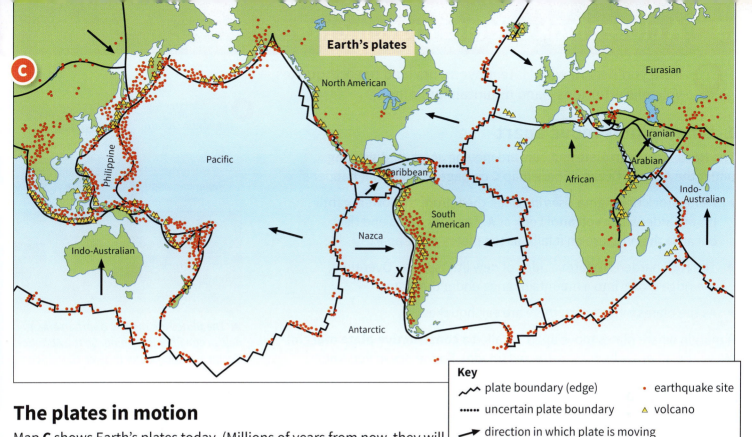

Earth's plates

North American
Eurasian
Pacific
Philippine
Caribbean
Iranian
Arabian
African
Indo-Australian
South American
Nazca
X
Indo-Australian
Antarctic

Key
- ∿∿ plate boundary (edge)
- •••••• uncertain plate boundary
- → direction in which plate is moving
- • earthquake site
- △ volcano

The plates in motion

Map **C** shows Earth's plates today. (Millions of years from now, they will look different.) Some carry continents and ocean, others just ocean. They move in different directions. Scientists use **GPS** to track them. They creep along at 2 – 15 cm a year!

Why do they move?

Today, scientists think that **gravity** is the main reason why plates move.

D shows three plates X, Y and Z. Follow the numbers. You can see that:

- plate Y is *pulled along* by its sinking edge. This is called **slab pull**.
- Y is also *pushed along* by the spreading ridge of lithosphere. This is called **ridge push**.

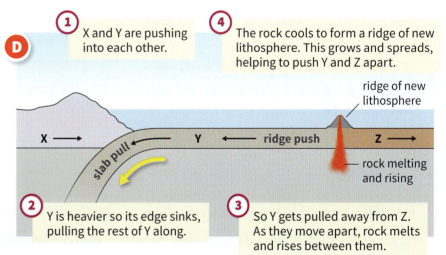

D

① X and Y are pushing into each other.

④ The rock cools to form a ridge of new lithosphere. This grows and spreads, helping to push Y and Z apart.

ridge of new lithosphere

X → slab pull ← Y ← ridge push Z →

rock melting and rising

② Y is heavier so its edge sinks, pulling the rest of Y along.

③ So Y gets pulled away from Z. As they move apart, rock melts and rises between them.

Your turn

1 Define: **a** earthquake **b** volcano **c** plate

2 Using map **C** to help you, name:
- **a** the plate you live on
- **b** a plate that is moving away from yours
- **c** a plate that is moving north
- **d** the largest plate, which carries just ocean
- **e** the plate off the west coast of South America
- **f** the plate circled by the Ring of Fire (*Did you know? …*)

3 Compare the patterns in maps **A** and **C**. Then use what you notice to *explain* the pattern in **A**.

4 Suggest a reason why the UK:
- **a** has no active volcanoes today
- **b** has the remains of several ancient volcanoes

5 What is a *plate margin*? Give another name for it.

6 Find X on map **C**, where two plates meet.
- **a** Name the two plates.
- **b** **i** Which of the two plates carries only ocean?
  - **ii** The edge of this plate sinks where the plates meet. Why?
  - **iii** Explain how *slab pull* keep this plate moving.
  - **iv** What is happening at the opposite edge of this plate?

Here you'll learn why plate movements lead to earthquakes, volcanic eruptions – and mountain building!

1 When plates move apart

The North American plate and ours are moving apart, or **diverging**, under the Atlantic Ocean. (Check map **C** on page 91.) So what happens?

- Melted rock or **magma** rises between them from the lower mantle. It's an underwater **volcano**! Look at **A**. The rock melts and rises because the pressure on it falls, as the plates move apart.

- The rock hardens to form a ridge of new lithosphere. Over time, the ridge grows into a mountain range under the water.

- As the plates move apart, there are earthquakes.

A margin where plates move apart is called a **constructive plate margin**. New lithosphere is being *constructed* (made). So the ocean gets wider!

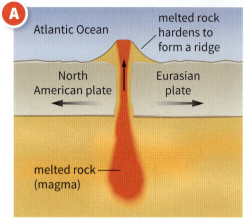

▲ The plates are moving apart. In effect, new ocean floor is forming. So the Atlantic Ocean is getting wider by over 2 cm a year.

2 When plates push into each other

The Nazca and South American plates are pushing into each other, or **converging**, just off the west coast of South America. (Map!)

- The Nazca plate is heavier. (Oceanic crust is heavier.) So its edge sinks at an **ocean trench**. Look at **B**.

- As it jolts downwards, there are earthquakes.

- It carries water with it, into the lower mantle. This lowers the melting point of the rock. So molten rock forces its way up through the Andes, to form a volcano.

- As the plates push into each other, the edge of the South American plate also crumples upwards. (That's how the Andes formed!)

A margin where plates push into each other is called a **destructive plate margin**. Ocean floor is being *destroyed*.

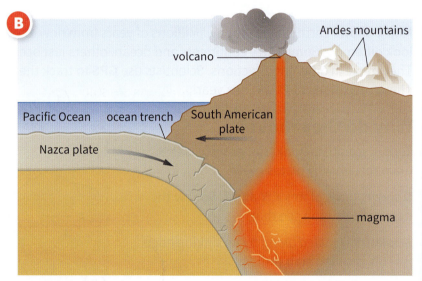

▲ The plates are pushing into each other. Ocean floor is being destroyed.

3 When plates slide past each other

For example the Pacific plate is sliding past the North American plate. (Map!)

- Both move in the same direction – but the Pacific plate is going faster.

- Sometimes they get stuck, then jolt free. This causes earthquakes.

- But no rock is melted, so there are no volcanoes. And no mountains form.

A margin where plates slide past each other is called a **conservative plate margin**. No lithosphere is being created, nor destroyed. It is kept as it is, or *conserved*.

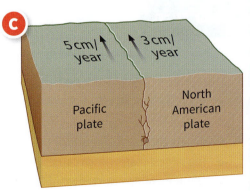

▲ Here, plates are sliding past each other. No lithosphere is created, or destroyed.

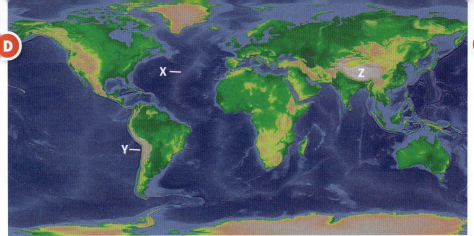

▲ A relief map of Earth. Look at the thin wavy ridges in the oceans. X is the Mid-Atlantic Ridge, where the North American plate and ours are moving apart. Magma erupts through the top of it. The dark line at Y is the Peru-Chile Trench, where the Nazca plate is sinking.

My deepest dive ... 10 925 m in the Mariana Trench.

▲ Safe journey! In 2018 – 19, the explorer Victor Vescovo dived to the bottom of the deepest trench in each ocean, in this submersible.

More about mountain building

As you saw in **2**, the Andes were formed by plates pushing into each other, under the ocean.

The Indo-Australian and Eurasian plates are also pushing into each other – but on land. (See map **C** on page 91.)

The two plates have the same density, so neither sinks. Instead, rock is squeezed upwards, forming the Himalayas. They are at Z on map **D**.

The plates are still pushing. So the Himalayas are still rising, by over 1 cm a year. There are many earthquakes.

The mountains of the Himalayas and Andes are called **fold mountains**. Can you see why?

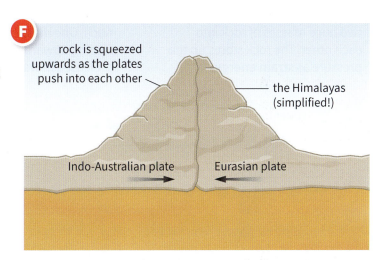

rock is squeezed upwards as the plates push into each other

the Himalayas (simplified!)

Indo-Australian plate   Eurasian plate

Your turn

1 a Make a table like this one, and complete it using ticks and crosses. A tick means *yes*.

Plate movement ...	... gives earthquakes	... gives volcanoes	... builds mountains
1 pulling apart			
2 pushing into each other			
3			

  b *All* plate movements give rise to …?

  c Fold mountains form where plates *greencov*. Unjumble!

2 Copy drawing **G** on the right. On your copy:

  a label the oceanic plate, continental plate, and a volcano

  b shade in, and label, melted rock that feeds the volcano

  c mark in and label any earthquake site

  d add a label and arrow for *slab pull*

  e add a title saying which type of plate margin this is

3 a What are *fold mountains*? (Glossary?)

  b Using the maps on pages 91 and 140 – 141, explain why there are fold mountains, volcanoes, and earthquakes in:
    i Peru   ii Italy

4 Explain why plate margins (boundaries) are called:

  a *destructive*, where plates are pushing into each other

  b *conservative*, where plates are sliding past each other

  c *constructive*, where plates are moving apart

5 Where plates move apart, the ocean floor gets wider. But Earth is not getting any bigger. Explain why.

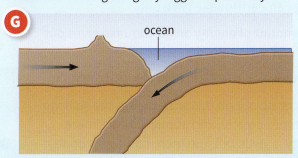

ocean

Earthquakes can kill. Here you'll learn what an earthquake is, how it is measured, and what damage it can do.

Did you know?
- Seismic waves tell us a lot about Earth's layers …
- … because they pass through different materials at different speeds.

It's an earthquake!

Let's see how an earthquake happens.

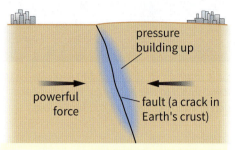

Imagine powerful forces pushing two huge masses of rock into each other. The rock stores up the pressure as **strain energy**.

pressure building up
powerful force
fault (a crack in Earth's crust)

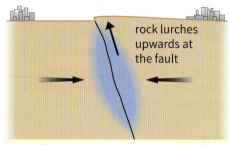

But suddenly, the pressure gets too much. One mass of rock gives way, slipping upwards. The stored energy is released in waves …

rock lurches upwards at the fault

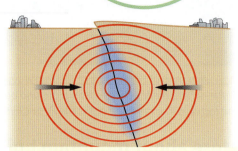

… called **seismic waves**. These pass through Earth in all directions, shaking everything. The shaking is called an **earthquake**.

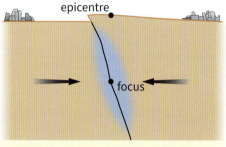

The **focus** of the earthquake is the point where the waves started. The **epicentre** is the point directly above it on Earth's surface.

epicentre
focus

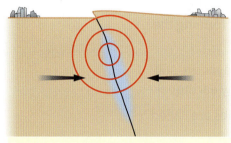

As the rock settles into its new position, there will be lots of smaller earthquakes called **aftershocks**.

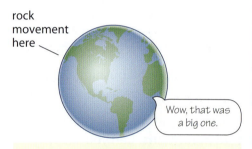

Seismic waves get weaker as they travel. Even so, a large earthquake can be detected thousands of kilometres away!

rock movement here
Wow, that was a big one.

Any sudden large rock movement can cause an earthquake. That's why there are so many earthquakes along plate edges. But even the collapse of an old mine shaft can cause a small earthquake.

How big?

- Earthquakes are measured using machines called **seismometers**. They record the shaking as waves on a graph.

- From the graph, scientists can tell how much energy an earthquake gave out.

- The amount of energy an earthquake gives out is called its **magnitude**.

- We show it on the **Richter scale**. (On the right.)

- An increase of 1 on this scale means the shaking is 10 times greater, and about 30 times more **energy** is given out. (So more damage is done.)

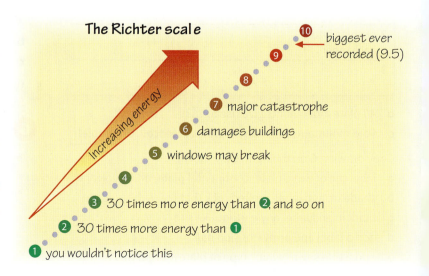

The Richter scale

Increasing energy

10 — biggest ever recorded (9.5)
9
8
7 — major catastrophe
6 — damages buildings
5 — windows may break
4
3 — 30 times more energy than 2, and so on
2 — 30 times more energy than 1
1 — you wouldn't notice this

What damage can an earthquake do?

An earthquake shakes the ground, which then shakes everything on it. So …

Earthquakes in the ocean floor can cause giant waves called **tsunami** (say *tsoonami*). They slam onto the land and destroy places.

Landslides may block roads.

Buildings and bridges crack and topple. Roads split open.

Water mains burst – which means no water.

Gas pipes fracture, and electricity cables are torn down. These cause fires.

Transport comes to a standstill.

At home, cookers and heaters fall over and start fires. Ceilings collapse. Doors jam. Everything slides off shelves and tables.

So earthquakes can destroy homes, villages, towns, and whole cities.
And the big problem is this: *scientists cannot predict them, and warn people*.

Your turn

1 Why do earthquakes happen so suddenly?

2 Explain in your own words what these earthquake terms mean. Use complete sentences.

  a seismic waves   b focus

  c epicentre   d magnitude

3 You are one of the people in the photo above. (Look carefully!) Describe what you see around you, in at least five lines.

4 Look at the earthquake diagram on the right.

  a Will the shaking be stronger at A, or at B? Explain.

  b Will the damage be greater at A, or at B? Why?

  c Will an earthquake of magnitude 7 do more damage than this one, or less? Why?

  d About how many times more energy will an earthquake of magnitude 7 give out, than this one?

  e An earthquake can occur at any time of day. When might an earthquake do more harm at B?

    i at 5 am   ii at 10.30 am

  Explain your answer.

5 The largest earthquake ever recorded was in 1960. Its epicentre was *in the ocean*, off the coast of Chile. It measured 9.5 on the Richter scale. Use the maps on pages 91 and 140 – 141 to help you answer these questions.

  a State the most likely cause of this earthquake. Include the names of two plates in your answer.

  b The earthquake killed over 1600 people in Chile. Suggest reasons for this death toll.

  c 22 hours later, it caused 200 people to drown on the east coast of Japan. How did it do that?

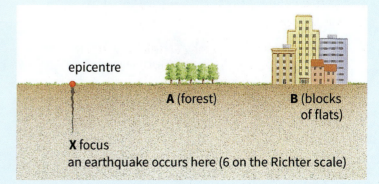

epicentre

A (forest)

B (blocks of flats)

X focus
an earthquake occurs here (6 on the Richter scale)

Here you'll learn why Southwest China is prone to earthquakes, and read about a deadly example.

CHINA
Equator

The Sichuan earthquake, 2008

A normal day in school

12 May 2008. It's a normal Monday afternoon in Beichuan Middle School, in Sichuan province, Southwest China. The 2900 students and teachers are deep in lessons.

But not for much longer. At 2.28 pm, over 140 km away, in the mountains, a large mass of rock starts to slip downwards. It sets off an earthquake.

Less than a minute later, Beichuan Middle School begins to shake. Books topple. Walls sway and crack. Door frames splinter. Ceilings crash down. Windows explode. People scream.

It seems to go on forever – but it lasts just two minutes. By then, two of the school buildings have collapsed, crushing students and teachers. In the days that follow, over 1300 bodies will be pulled from the rubble.

They are not alone. A third of Beichuan's population of 160 000 lie dead. 80 % of its buildings have been destroyed. 10 000 people are injured.

Not just Beichuan

Villages, towns and cities across a large area of Sichuan are shaken. The death toll is over 87 000. Of these, 5600 are students who die in class. Over 12.5 million farm animals are killed.

Earthquake-prone

China has many earthquakes – and many are in the southwest. In the hundred years up to 2008, earthquakes caused over 650 000 deaths in China. The 2008 earthquake is the costliest in financial terms.

▲ *A student waits to be rescued, in Beichuan.*

▲ *Outside a school in Beichuan, after the earthquake.*

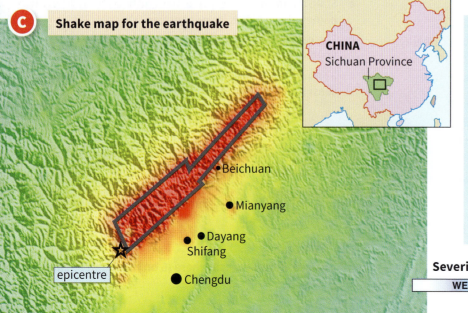

Shake map for the earthquake

CHINA
Sichuan Province

Beichuan
Mianyang
Dayang
Shifang
epicentre
Chengdu
Ziyang

0   50   100 km

Sichuan 2008: earthquake factfile

date	Monday 12 May 2008
time	2.28 pm
magnitude	7.9 on the Richter scale
epicentre	in mountains in Sichuan Province, Southwest China
people	over 87 000 dead, over 370 000 injured, at least 5 million left homeless
financial cost	nearly £120 billion

Severity of shaking

WEAK   STRONG   SEVERE

What caused the earthquake?

Like most big earthquakes, this one was caused by plate movements.

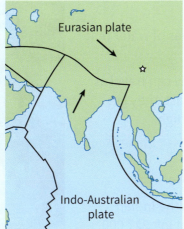

① The Indo-Australian plate is pushing into the Eurasian plate. (Map **C** on page 91.)

② The strain has caused many cracks or **faults** to develop in rock around the plate edges.

Eurasian plate

Indo-Australian plate

③ On 12 May 2008, the strain got too much at a fault in Sichuan. Rock slipped downwards along the fault, setting off the earthquake. (☆ marks the location of the epicentre.)

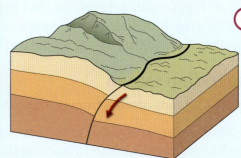

④ In places, the rock dropped by as much as 12 metres.

Why were so many killed?

Earthquakes don't kill. Buildings do. Many buildings could not withstand the shaking. They collapsed, crushing people. Many were schools.

After the earthquake

Help poured in from the Chinese government. Soldiers led the rescue effort. Over the next few days, many people were pulled from the rubble, still alive. But then came the time when only dead bodies were found.

Later, parents blamed local officials for the deaths of their children. They said schools had been badly built, because money intended for schools had been stolen, or used for other things. It became a national scandal.

Today

Today, weeds grow through the ruins of Beichuan. Most of the survivors live in a new town, about 30 km away. All the new buildings have been designed to withstand earthquakes. Because one thing is certain: as long as those plates move, there will be earthquakes. In Sichuan, and other parts of China.

▲ *Rescued from the rubble in Mianyang.*

Your turn

1 Where is Sichuan? Give the continent *and* country.

2 Imagine you are the student in photo **A**. Describe what happened in your classroom during the earthquake … and then, how you were rescued.

3 What caused the earthquake? Explain in five lines. (**D**?)

4 **a** Look at map **C**. Why is it called a *shake map*?

  b How can you tell from **C** that the earthquake occurred in a mountainous area?

5 **a** What does the red colour on shake map **C** represent?

  b The cities of Mianyang and Ziyang are similar in size. Mianyang suffered much more than Ziyang did. Why?

  c Severe shaking was felt a long way from the epicentre. About how far away, at most? Use the scale.

6 There will be more earthquakes in Sichuan. Explain why.

7 The earthquake led to over 87 000 deaths. Could they have been prevented? Think, decide, and then explain your answer.

Earthquakes occur in the ocean floor too, causing tsunami.
This is about a deadly tsunami that travelled the Indian Ocean.

What is a tsunami?

Imagine a big earthquake in the ocean floor. It will set off waves of water that travel in all directions. These waves are called a **tsunami**.

Out in the ocean the waves may be only a metre high. But they can travel at over 700 km an hour. As they reach shallower water near a coast, they get slower, and taller. When they hit land they can be up to 30 m high.

A tsunami in the Indian Ocean

Tsunamis can kill people thousands of miles from the epicentre.

On 26 December 2004 there was an earthquake in the floor of the Indian Ocean. Huge: 9.2 on the Richter scale. It caused a tsunami that left 227 000 people dead, and hundreds of thousands homeless, *across fifteen countries*. It travelled nearly 7000 km across the ocean. Follow the numbers in order.

▲ *People flee as the tsunami slams onto Raya island in Thailand.*

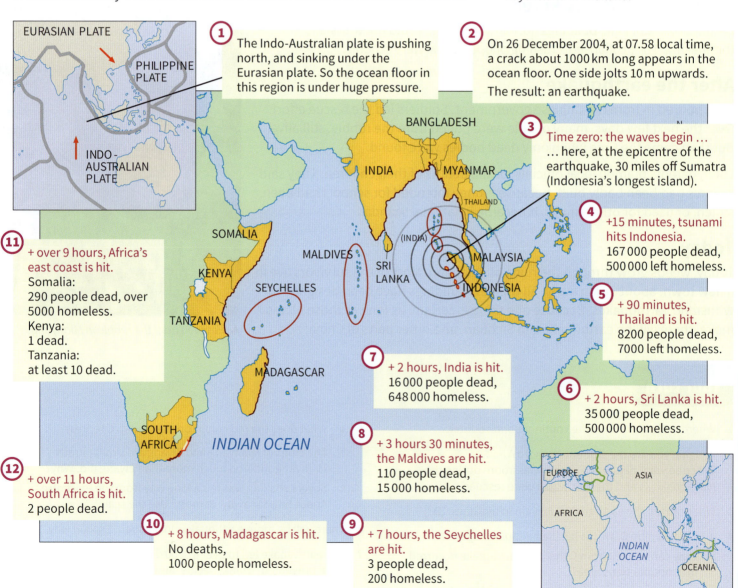

1 The Indo-Australian plate is pushing north, and sinking under the Eurasian plate. So the ocean floor in this region is under huge pressure.

2 On 26 December 2004, at 07.58 local time, a crack about 1000 km long appears in the ocean floor. One side jolts 10 m upwards. The result: an earthquake.

3 Time zero: the waves begin …
… here, at the epicentre of the earthquake, 30 miles off Sumatra (Indonesia's longest island).

4 +15 minutes, tsunami hits Indonesia. 167 000 people dead, 500 000 left homeless.

5 + 90 minutes, Thailand is hit. 8200 people dead, 7000 left homeless.

6 + 2 hours, Sri Lanka is hit. 35 000 people dead, 500 000 homeless.

7 + 2 hours, India is hit. 16 000 people dead, 648 000 homeless.

8 + 3 hours 30 minutes, the Maldives are hit. 110 people dead, 15 000 homeless.

9 + 7 hours, the Seychelles are hit. 3 people dead, 200 homeless.

10 + 8 hours, Madagascar is hit. No deaths, 1000 people homeless.

11 + over 9 hours, Africa's east coast is hit. Somalia: 290 people dead, over 5000 homeless. Kenya: 1 dead. Tanzania: at least 10 dead.

12 + over 11 hours, South Africa is hit. 2 people dead.

▲ As a tsunami nears the coast, it sucks up water, exposing the ocean floor. This satellite image shows the water being dragged away, at the resort of Kalutara in Sri Lanka.

▲ The same resort, after the tsunami has struck. The water churns and recedes, leaving destruction behind. Both images were taken on 26 December 2004.

The day they will never forget

Banda Aceh, Indonesia: I took an early ferry. I thought it was bouncing a bit, but that did not worry me. After an hour we got to Banda Aceh. I could not believe my eyes. The fishermen's homes along the water had gone. In the town, there were fishing boats on roof tops, and taxis stuck in trees. People were sobbing. There were bodies everywhere.

Telwatta, Sri Lanka: I was on the coast train, going see my family. Suddenly the train stopped. The sea started to pour in, very fast. The train turned over and over. I was trapped in there for nearly an hour, half drowned. But I'm lucky. They say there were 1500 passengers, and 800 of them died.

Khao Lak, Thailand: There was a hissing noise, and all the water along the beach got sucked out to sea. Lots of fish were left flapping on the ground. Children ran to look at them. Then there was a noise like thunder, and we saw a giant wave coming. The children had no chance.

(Adapted from news reports, December 2004)

▲ Three days after the tsunami, on the island of Phi Phi in Thailand.

Did you know?
- A 10-year old girl from the UK saved over 100 lives in Thailand, in the 2004 tsunami.
- She'd learned about tsunamis in geography class, and told people to run from the beach.

A warning system

Because of the 2004 tsunami, there is now a tsunami warning system for the countries around the Indian Ocean. When an earthquake in the ocean is detected, people are warned through sirens and loudspeakers.

Your turn

1 What causes a tsunami?

2 Explain these facts about the 2004 tsunami.
   a It reached more than a dozen countries.
   b It arrived at each country at a different time.
   c Indonesia suffered much more than Somalia did.
   d People out at sea were not aware of the tsunami.
   e It did not reach the Philippines. (Page 141?)

3 Tsunamis can do far more harm than earthquakes do. Explain why.

4 Could a tsunami strike the UK? Check the map on page 141, then decide and give your reasons.

5 A tsunami warning system includes a set of buoys that float on the water and give out signals. Suggest a reason why these buoys are a very important part of the system.

Here you'll learn about volcanoes, and the damage an eruption can do.

Here comes liquid rock!

A **volcano** is where liquid rock or **magma** shoots out, or **erupts**, through the ground. Above ground, the liquid rock is called **lava**.

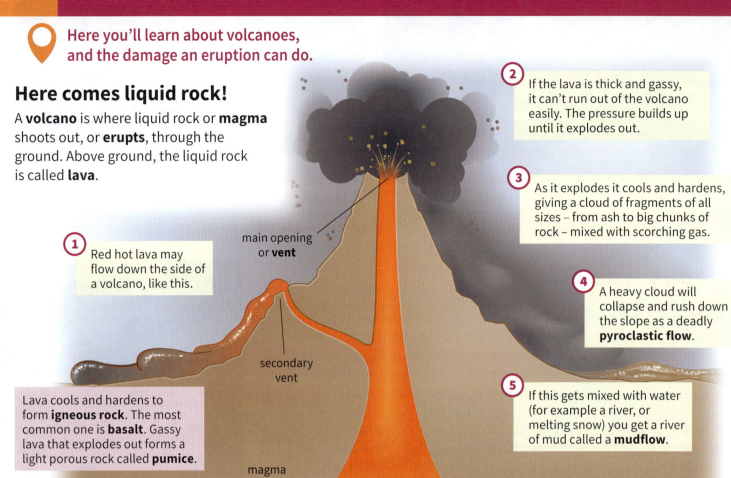

2 If the lava is thick and gassy, it can't run out of the volcano easily. The pressure builds up until it explodes out.

3 As it explodes it cools and hardens, giving a cloud of fragments of all sizes – from ash to big chunks of rock – mixed with scorching gas.

4 A heavy cloud will collapse and rush down the slope as a deadly **pyroclastic flow**.

1 Red hot lava may flow down the side of a volcano, like this.

main opening or **vent**

secondary vent

Lava cools and hardens to form **igneous rock**. The most common one is **basalt**. Gassy lava that explodes out forms a light porous rock called **pumice**.

magma chamber

magma (the melted rock mixed with gas)

5 If this gets mixed with water (for example a river, or melting snow) you get a river of mud called a **mudflow**.

Magma can be **viscous** (thick like tar) or as runny as thin custard. It depends on the rock that melted.

Volcanic gas is mainly steam and carbon dioxide, plus some sulphur dioxide and other gases. It smells of rotten eggs. It can suffocate you.

Viscous gassy lava is the most dangerous kind. It builds up inside the volcano. Then the gas propels it out in an explosion.

A

▲ *Runny lava flowing into the Pacific Ocean in Hawaii, USA.*

B

▲ *Ash erupting from a vent in a volcano in Guatemala. The hollow around a vent is called a **crater**. The vent on the right is inactive.*

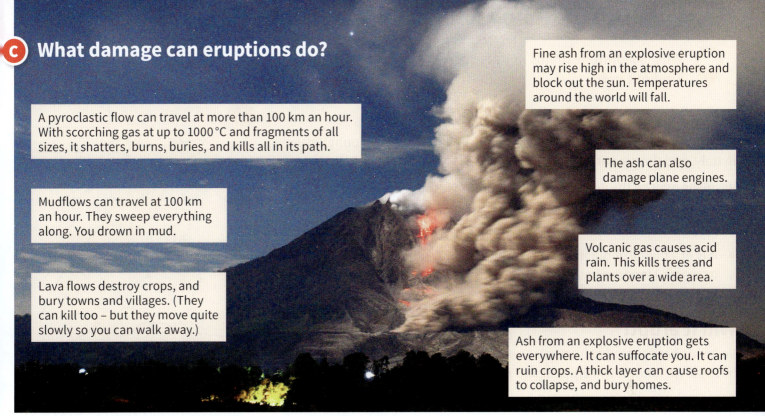

C What damage can eruptions do?

Fine ash from an explosive eruption may rise high in the atmosphere and block out the sun. Temperatures around the world will fall.

A pyroclastic flow can travel at more than 100 km an hour. With scorching gas at up to 1000 °C and fragments of all sizes, it shatters, burns, buries, and kills all in its path.

The ash can also damage plane engines.

Mudflows can travel at 100 km an hour. They sweep everything along. You drown in mud.

Volcanic gas causes acid rain. This kills trees and plants over a wide area.

Lava flows destroy crops, and bury towns and villages. (They can kill too – but they move quite slowly so you can walk away.)

Ash from an explosive eruption gets everywhere. It can suffocate you. It can ruin crops. A thick layer can cause roofs to collapse, and bury homes.

Photo **C** shows an eruption of Mount Sinabung in Indonesia. After more than 1200 years of inactivity, this volcano erupted in 2010, and then several times since.

By 2019, its eruptions had killed 25 people in the small villages on its slopes. Luckily, volcanoes show signs that they will soon erupt. For example gas emissions, and swarms of small earthquakes caused by magma on the move. So people were warned, and most were able to move to safety.

▲ Mt Sinabung is on Sumatra, in Indonesia.

Your turn

1 What is the difference between *magma* and *lava*?

2 Make a larger copy of this drawing of a volcano. Colour in your copy, and add the missing labels.

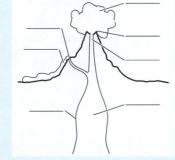

3 An active volcano can produce:
   ash   a pyroclastic flow   a lava flow   volcanic gases

   a List these in order of danger for the local people, starting with what you think is the most dangerous.

   b Beside each item in your list, say what harm it does.

   c Which items in your list could affect other countries? Underline them, and explain your choice.

4 Photo **D** below shows rescuers after an eruption of Mount Sinabung in 2014. (It killed 16 people.) Imagine you are one of the rescuers. Describe what you see around you.

5 Most of the people around Mount Sinabung live by farming. It could be many years before people return to live in the area in **D**. Suggest at least two reasons.

6 Mount Sinabung is likely to continue erupting. Suggest at least two ways in which the death toll from its eruptions could be kept to a minimum. (More than two if you can!)

D

Why is Mount Vesuvius in Italy such a dangerous volcano? Find out here.

Mount Vesuvius

Look at **B** below. It is a false-colour satellite image. It shows Mount Vesuvius in Italy, and the area around it. Check out its location on map **A**.

Vesuvius may be the world's most dangerous volcano. Because …

- it is fed by thick, sticky, gassy magma – the kind that leads to explosive eruptions, and deadly pyroclastic flows.

- it is in a highly populated area. Around 600 000 people live within 8 km of its crater. Many are in shoddy homes built on its slopes illegally. Around 3 million people live within 30 km.

Will it erupt?

Vesuvius has erupted at intervals over thousands of years. The last time was in 1944. We know it will erupt again. The question is: *When?*

So Vesuvius is monitored 24/7. It has instruments to detect swelling on its slopes, earthquakes due to magma on the move, and gas emissions. Satellites also monitor it for changes, including changes in temperature.

Even so, scientists can give only three days' warning of an eruption, at most. That is not much time to **evacuate** everyone from the danger zone.

A

Key
— plate margin
▲ volcano

▲ The African plate is sinking under the Eurasian plate, and this gives rise to Italy's volcanoes. Campi Flegrei ('burning fields') is a complex volcano, close to Naples.

Key
▨ built up area
▨ open area, showing relief

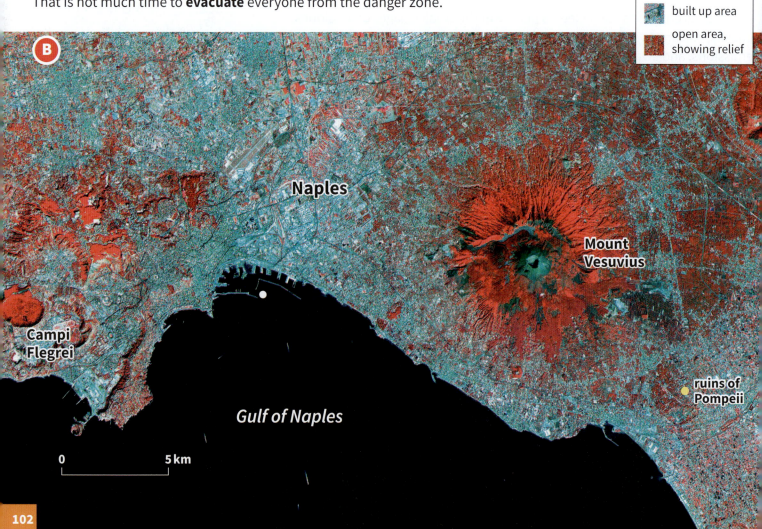

B

Naples

Mount Vesuvius

Campi Flegrei

ruins of Pompeii

Gulf of Naples

0   5 km

The eruption that destroyed Pompeii

In CE 79, Pompeii was a small wealthy Roman city.

One day in autumn that year, around 1 pm, Vesuvius erupted. A boiling column of gas and ash and lava shot over 30 km into the air. As it collapsed, pyroclastic flows raced down the slopes, destroying everything.

The eruption lasted for two days. When it was all over, Pompeii lay buried under 5 metres of ash and pumice. Its ruins are marked on **B**.

Finding Pompeii

For centuries, Pompeii lay forgotten.

The first major excavations began in the 18th century. Archaeologists found hollows where bodies had decayed in the ash. By filling the hollows with plaster, they could reconstruct the last hours of Pompeii.

Excavations are still not complete. But Pompeii is now a major tourist attraction.

▲ Plaster casts of over 1100 bodies have been created in Pompeii. People were buried under hot choking ash and cinders of pumice.

Evacuation plans

People will be told to evacuate, when Vesuvius shows signs of erupting.

First to go will be the 600 000 people living closest to Vesuvius. Others will follow, depending on the risk.

They will be moved to other parts of Italy by rail, road, and boat.

They may have to stay away. The area may not be habitable for many years.

But people have many concerns about the plans, and planners. Look at **D**.

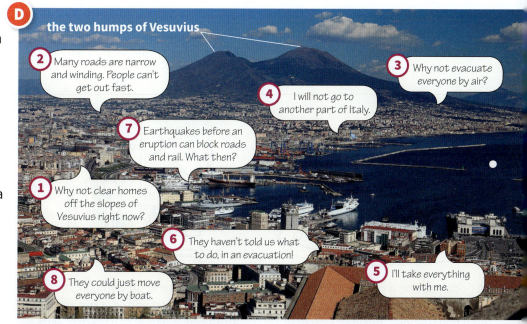

the two humps of Vesuvius

2 Many roads are narrow and winding. People can't get out fast.

3 Why not evacuate everyone by air?

4 I will not go to another part of Italy.

7 Earthquakes before an eruption can block roads and rail. What then?

1 Why not clear homes off the slopes of Vesuvius right now?

6 They haven't told us what to do, in an evacuation!

5 I'll take everything with me.

8 They could just move everyone by boat.

▲ View of Vesuvius from Naples. Find the matching white dots in the sea in **D** and **B**.

Your turn

1 Where is Mount Vesuvius? Name the continent, country, and the nearby city and sea area.

2 Give two reasons why Mount Vesuvius is very dangerous.

3 a Campi Flegrei is a *complex volcano*. Explain. (Glossary?)
   b Using **A** to help you, explain why Italy has volcanoes.

4 Pompeii was destroyed when Vesuvius erupted in 79 CE. How far are its ruins from the volcano's *crater*? (Glossary.)
   about 2 km   about 8 km   about 20 km

5 Vesuvius is monitored closely. Explain why, in not more than six lines. (Include the word *evacuate*.)

6 Vesuvius will erupt again, one day. So planning is essential. This is about the comments and suggestions in **D**.

   a Is suggestion 1 a good idea? Evaluate it, and explain.

   b Is suggestion 3 a good idea? Explain your answer.

   c Choose a different suggestion from **D** that you think is sensible, and support your choice.

   d Choose five comments from **D** which are about things that could turn an evacuation into chaos.

   e Identify two things the planners could do straight away, to help a future evacuation go smoothly.

Plate margins are dangerous places. But people still live along, or close to, them. Why? Find out here.

What if ...
... the UK were in a danger zone?

Dangerous living?

Many millions of people live near plate margins, where the risk of earthquakes, and eruptions, is high. Why don't they move somewhere safer?

What danger?

Los Angeles 1 Mile

People settled in danger zones before we understood the risks. (We didn't know about Earth's plates until the 1960's.)

We have to live somewhere.

Some settlements in danger zones are enormous cities. Mexico City and Tokyo, for example. Where would everyone go?

This is our way of life.

Even after a disaster, most people want to return to the life they know best. They hope another disaster won't happen.

Life is good here.

A good job and a pleasant life may keep you in a danger zone, even if you feel nervous. Besides, you may feel safer because ...

Right on time.

... scientists monitor volcanoes, and are getting better at predicting eruptions. So they can warn you to move to safety.

So it will bend, and flex ...

... and now I'll cushion it with rubber.

They can't predict earthquakes ... yet. But engineers design quake-proof buildings, bridges, and other structures that won't collapse.

▼ *Yokohama, Japan. On the left is the Landmark Tower (●). It is built on rollers, and from flexible materials, to withstand earthquakes.*

▼ *Lemons from a citrus grove at the foot of Mount Etna, an active volcano in Sicily, Italy. That's Mount Etna in the background.*

A

B

Now for the good news!

You've met the bad news: plate movements can kill.
The good news is: they also bring benefits – thanks mainly to volcanoes.

Good soil. The lava from volcanoes breaks down to give very fertile soil. On Mount Etna in Sicily, for example, the soil gives farmers rich crops of grapes and other fruit. (Photo **B**.)

Money from tourism. Volcanic areas attract tourists, and tourists spend money! They flock to visit Mount Fuji in Japan, and the volcanoes of Italy and Iceland. They can see **geysers** and **fumaroles**, and relax in hot springs.

Geothermal energy, or heat energy stored in rock. For example, Iceland is a volcanic island. Water is pumped down into the hot rocks. It comes back up as steam. This is used to heat homes, or drive turbines to make electricity.

Over 90 % of Iceland's homes are heated this way.

Valuable materials. Copper, silver, gold, and lead are found in extinct volcanoes. (They collect in **veins** when magma cools and hardens.) Sulphur is mined around old volcano vents. Basalt is used to build roads.

Fossil fuels. Oil and gas form deep in the ocean floor, from the soft remains of sea creatures. Plate movements caused uplifting of ocean floor, giving us oil and gas wells on land. They also led to the burial of vegetation, giving us coal.

Fossil fuels have benefited us. But now the world is turning from them, because they are linked to climate change.

C

▲ *A geothermal power station in Iceland. Steam is made using the hot rock below. Iceland lies on the Mid-Atlantic Ridge – and it is being split as the plates move apart!*

Your turn

1 Using the world map on pages 140 – 141, and **C** on page 91, name six capital cities which appear to be in danger zones – at or close to plate margins.

2 Give: **a** two economic reasons **b** two social reasons to explain why people continue to live in danger zones.

3 Japan has more than 1500 earthquakes a year.
a Why does it have so many earthquakes? (Check the maps on pages 91 and 141.)
b Look at the Yokohama Landmark Tower in photo **A**. Explain how rollers and flexible materials help to make it quake-proof.

4 Do you agree with this person? Decide, and then justify your decision.

MOVE EVERYONE OUT OF DANGER ZONES **NOW!**

5 Define: **a** fumarole **b** geyser (Glossary?)

6 **a** What is *geothermal energy*?
b The power station in **C** does *not* contribute to climate change. Explain why.
c Why is geothermal energy so widely available in Iceland?

7 In what ways have plate movements benefited *you*? Give your answer as a spider map.

8 Some people do benefit from earthquakes. Who? Give as many examples as you can.

9 You are an engineer. Copy this drawing and complete it to show how you could heat those homes. Add a title!

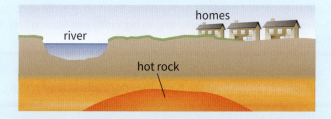

homes

river

hot rock

5 **Our restless planet**
How much have you learned about our restless planet? Let's see.

check ✓

1 **A** shows a slice through Earth. One layer, the crust, is named. P and Q are Earth's two other layers.

 a State the name of the layer labelled: **i** P **ii** Q

 b Which of the three layers contains the lightest rock?

 c **i** What does label R represent?

  ii The rock in R is broken into big rigid slabs. These are called .... ?

 d *Rigid* means stiff, not flexible. The rock at S is solid – but it is not rigid. How does it behave?

 e Name the type of crust that is labelled: **i** T **ii** U

 f Which type of crust is heavier, T or U?

2 **B** shows the plate movements that created Mount Vesuvius, and Italy's other volcanoes.

 a What are these plates doing: *converging*, or *diverging*?

 b The African plate is sinking below the Eurasian plate. Why?

 c Where does the magma come from, that feeds Vesuvius?

 d **i** Scientists monitor Vesuvius very closely. What do they want to find out, and why?

  ii Satellites monitor the temperature of Vesuvius. It will rise when an eruption is on the way. Explain why.

 e Italy experiences many earthquakes. Give a reason.

 f Unlike Italy, the UK has no active volcanoes. Why not?

3 Look at map **C.** The British Isles are circled.

 a X is a long mountain ridge under the ocean. State its name.

 b The ridge shows where plates meet. Which two plates meet along X in the northern hemisphere?

 c Explain why these occur along the ridge:
  i volcanic eruptions **ii** earthquakes

 d Name the type of rock the ridge is made of.

4 **D** shows the scene after an earthquake in China.

 a Earthquakes can occur anywhere, even in the UK. (They are usually weak in the UK.) What causes them?

 b Explain why many earthquakes occur along and near plate margins.

 c Name the waves of energy that earthquakes give out.

 d **i** What scale is used to compare the strength of earthquakes?

  ii Which of these numbers on the scale shows a more powerful earthquake? 3.5 9.2

 e Give one reason why earthquakes can lead to many deaths. **D** will give you a clue.

 f What can be done to reduce earthquake deaths, for people living in places like **D**?

5 Scientists cannot predict earthquakes. Even so, people continue to live near plate margins, including in big cities like Mexico City and Tokyo. Give two reasons why they remain.

6 *Earthquakes and volcanic eruptions cause deaths and damage.* To what extent can we limit the harm they do? Write at least half a page.

A

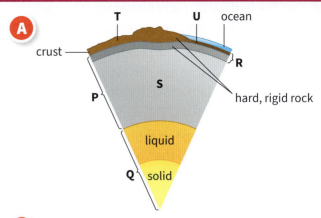

crust — T U ocean
P S hard, rigid rock
liquid
Q solid

B

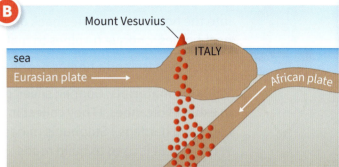

Mount Vesuvius
ITALY
sea
Eurasian plate →
African plate

C

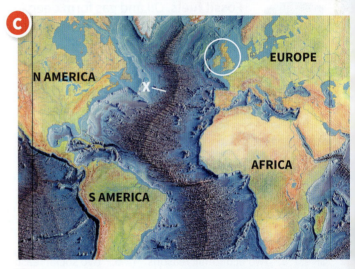

N AMERICA EUROPE
X
AFRICA
S AMERICA

D

6 About Russia

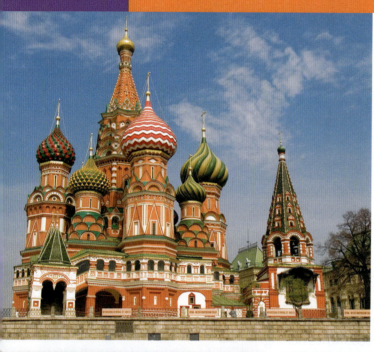

This unit introduces Russia – a big country that's often in the news.

Where is it?

When you are studying a new country, the first thing to do is find out where it is! So … where is Russia?

- It is in the **Northern Hemisphere**. It spreads across the top half of **Asia**, and into **Europe**. Look at **A**.

- It is bordered by two oceans.

- It is only 82 km from Alaska in the United States, across a strip of sea called the **Bering Strait**.

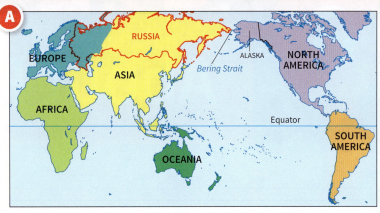

▲ The red line marks Russia's border. Alaska, in the USA, was once a Russian colony. The USA bought it from Russia in 1867, for $7.2 million.

Who are its neighbours?

Look at map **B**. Russia shares a land border with fourteen countries.

Note the small Russian **exclave** west of Lithuania. It is called **Kaliningrad**.

Look at the countries in pale pink. For much of the 20th century, these were part of the **Union of Soviet Socialist Republics** (the **USSR** or **Soviet Union**). This was dominated by Russia, and run from Moscow. But it broke up in 1991.

Now find **Crimea**. Under international law, it is part of Ukraine. But Russia controls it. In 2014, residents – supported by Russia – voted for it to become part of Russia. The vote was condemned as **illegal** by Ukraine and other countries, including the UK. The United Nations – a body that almost all countries belong to – does not accept that Crimea is part of Russia.

Did you know?

- Many countries imposed sanctions on Russia over its actions in Crimea.

- These include bans on some exports to Russia.

Key

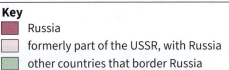

■	Russia
■	formerly part of the USSR, with Russia
■	other countries that border Russia

Compare!

What's it like?

Here is a quick overview …

- **Size** Russia is vast! It is the world's biggest country: 17.1 million sq km. About 70 times bigger than the UK. Almost twice the size of the USA.

- **Width** Russia is so wide that it has 11 time zones. At midday in Kaliningrad, it's 10 pm on the Kamchatka peninsula! (Find these on map **B**.)

- **Population** 146 million people – about 2.3 times as many as in the UK.

- **Empty in places** Huge areas are too cold for people to live in.

- **Landscapes** Frozen tundra, vast forests, grassy plains, sunny beaches, mountains, volcanoes, and long, long rivers. Russia has them all.

- **Natural wealth** Russia is rich in natural resources: oil, gas, coal, timber, metals, diamonds. It earns money from exporting these. Look at **C**.

How did it get so big?

Humans create countries. So how did Russia get so big?

- It began about 1300 CE, as a small state called **Muscovy**, about the size of Wales. This was centred on Moscow, and ruled by princes.

- It expanded by taking control of the peoples and lands around it. Just as the Roman Empire, a thousand years earlier, had spread from Rome.

- In 1721 the expanded territory was named **the Russian Empire**, with Peter the Great as **Tsar** (supreme ruler). It had many different ethnic groups.

- The empire continued to expand. But in 1917, **the Russian Revolution** forced the last tsar, Nicholas II, to step down. The empire was now a **republic**.

- The next leader was Lenin. He promoted **communism**, where everyone would be equal. A huge change! (Under the tsars, most people were very poor peasants.)

- Under Lenin, the territory was reorganised to form the Union of Soviet Socialist Republics. The larger non-Russian ethnic groups now had their own republics – but controlled by Moscow and the Communist Party.

- In 1991 the USSR voted to break up. The Communist Party no longer had power. Russia stood alone – but still vast.

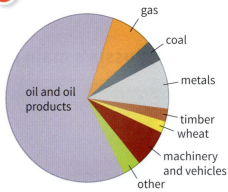

C Russia's exports in 2020

▲ Russia earned around $335 billion from these exports in 2020. Exports in 2019 earned $420 billion. Can you think of a key reason for the decline in 2020?

▲ Russia has lots of brown bears. They look cuddly – but watch out! The brown bear is often used as a symbol for Russia.

Your turn

1 a Russia lies on two continents. Which two?
   b Which continent has the bigger share of Russia?

2 a Name:
      i three Asian countries   ii four European countries
      which border Russia. Map **A** will help.
   b Which country shares the *longest* border with Russia?

3 Name two oceans which border Russia.

4 a Kaliningrad is an exclave. Define *exclave*. (Glossary?)
   b Crimea is *not* generally accepted as part of Russia. Why?

5 a How far is Russia from the USA, at their closest point?
   b Name the stretch of water that separates them.

6 About how much of Russia lies north of the Arctic Circle?
   a half   b one-tenth   c one-fifth

7 Explain why Moscow controlled:
   a a far larger area in 1721 than in 1300
   b a smaller area in 2020 than in 1990

8 *Most of Russia's exports are primary products*. True or false? Support your answer with evidence from **C**. (Glossary?)

When you study a new country, you need to get an idea of its physical geography. Read on!

A map of Russia's main physical features

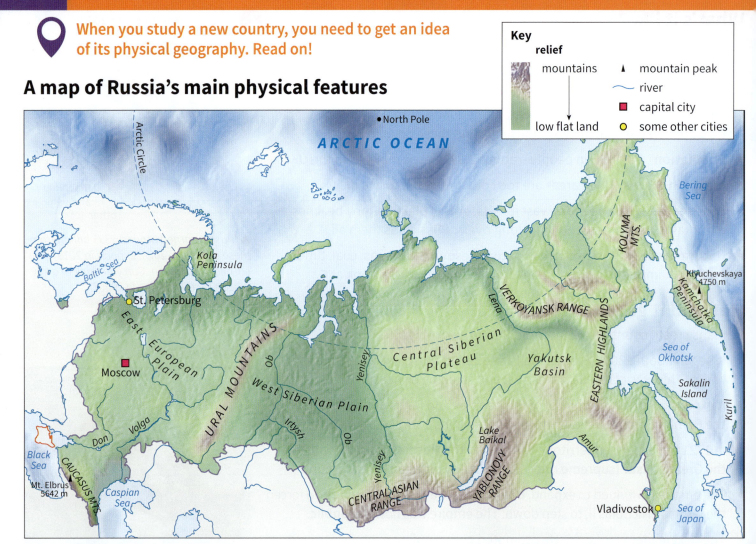

Key

relief

mountains ▲ mountain peak

low flat land

～ river

■ capital city

○ some other cities

Find these mountains …

- The **Ural Mountains**. These form a natural border between Asia and Europe. Asian Russia – to their east – is called **Siberia**.

- The **Caucasus Mountains**. Their highest peak is Mount Elbrus. At 5642 m, it is Europe's highest mountain.

Find these plains and plateaus …

Russia's plains and plateaus cover vast areas.

- The **East European Plain** is where most of Russia's population lives. It has the best farmland, and a less harsh climate overall.

- The **West Siberian Plain** is low and flat. It has lots of boggy land, and coniferous forests, and frozen tundra north of the Arctic Circle. Under this plain lies a massive basin of rock, that contains much of Russia's oil and gas. It covers over 2 million square km!

- The **Central Siberian Plateau** is an upland area, flat in places. Most of it is covered in coniferous forest. It is rich in resources: coal, oil and gas, nickel, iron, copper, gold, silver, platinum, diamonds, and more.

▲ Mount Elbrus, Europe's highest mountain. It's a volcano with two peaks, in the Caucasus mountain range. It last erupted nearly 2000 years ago – but still gives off fumes.

Look for these lakes and seas …

- **Lake Baikal**. It is the world's deepest lake. It holds about *one-fifth* of Earth's liquid fresh water. Think about that!

- The **Caspian Sea.** Salty, like a sea. But in fact it's a huge lake – the world's largest in area. Over 130 rivers flow into it, including the Volga. None flow out! But it does not overflow, because it loses water by evaporation.

- The **Baltic Sea**. It joins the North Sea (which borders Britain).

- The **Black Sea**. It flows into the Mediterranean Sea. The climate is mild, with warm sunny summers. So you'll find Russia's seaside resorts here.

And these rivers …

- The **Volga** is the longest river in Europe. It has a special place in Russian history and culture.

- Now look at the three great rivers of Siberia, that flow north for thousands of kilometres to the Arctic Ocean:
  - the **Ob**; the **Irtysh** is its main tributary
  - the **Yenisey**
  - the **Lena**.

  These rivers meander across huge flood plains. They often flood, because their upper reaches (in the south) thaw first after winter. Then, on its way north, the water gets dammed up behind ice.

- The **Amur** forms a natural border between Russia and China.

Volcanoes too!

- The **Kamchatka Peninsula** and **Kuril Islands** have many active volcanoes. They are part of the **Ring of Fire** around the Pacific Ocean (page 90).

- Mount Elbrus is also a volcano. It last erupted about 2000 years ago. (It formed when plate movements closed up an ancient ocean, and folded land upwards to form the Caucasus Mountains.)

Did you know?
- Russia has over 2.8 million lakes!
- 98 % of them are less than 1 sq km in area.

▲ *Out for a run on Lake Baikal in winter.*

▲ *A fuel tanker on the River Volga. The river is used a lot for transport.*

▲ *The tallest volcano in Kamchatka erupting: Mount Klyuchevskaya (4750 m).*

Your turn

1 Which of Russia's physical features …
  a acts as the border between Europe and Asia?
  b contains Europe's highest peak?
  c sits on a vast basin of rock that holds oil and gas?
  d lies between the Bering Sea and the Sea of Okhotsk?
  e contains about 20 % of the world's liquid fresh water?

2 What and where is *Siberia*?

3 State three facts about the Caspian Sea.

4 a Which river forms a border between Russia and China?
  b i Name five other Russian rivers, from the map.
    ii Name one river that lies completely in Europe.

5 Flooding is common on the River Lena as summer approaches, even when there's not much rain. Explain why.

6 Now … start a spider map to summarise what you know about Russia. Use a big sheet of paper, so that you can add more points in later lessons. Choose suitable headings.

Climates and biomes are closely connected. Here you can explore the link between them, for Russia.

Russia's climate zones

You know already that …

- **latitude** and **altitude** influence the climate.

- **distance from the sea** is a big factor too. Land further from the sea is warmer in summer, and cooler in winter.

Look at map **A**. How well does it match those factors?

tundra	long bitterly cold dark winters, short cool summers, precipitation low
sub-arctic	long very cold winters, short cool summers, precipitation low
humid continental	cold winters, warm/hot summers, precipitation higher (most in summer)
semi-arid	cold winters, hot summers, dry
mountain	the higher you go, the colder it gets

Cold, cold, cold …

Much of Russia is under snow *for up to eight months of the year*. The tundra and sub-Arctic climate zones have **permafrost** – frozen soil below the ground. It can be over 1 km thick! Only the surface layer thaws in summer.

Precipitation is low, overall. That's mainly because in cold places there is little evaporation, so the air is dry. Much of Russia is also sheltered by mountains, which block moist air. Rain falls mostly in summer.

Russia's biomes

Biomes are regions with similar natural vegetation, and animals. Map **B** is a simplified map of Russia's biomes. The boxes will tell you more.

▲ *An Arctic fox in the tundra, still in its winter coat. Its summer coat is brown.*

Key			
☐	tundra	☐	temperate forest
☐	taiga	☐	mountain (also called Alpine)
☐	steppe		

Tundra

The **tundra** biome is covered in ice and snow in winter. But in summer the surface of the soil thaws, and low plants and shrubs grow. You may see musk ox, Arctic foxes, polar bears, brown bears, reindeer, and ermine.

Taiga

The **taiga** is a biome of *coniferous* forest – trees like larch, spruce and pine, with needles instead of leaves. Larch has shallow roots, so is common in permafrost. Watch out for bears, wolves, and Siberian tigers!

It's winter, so the river is frozen.

Temperate forest

Here you'll find a mixture of trees: *deciduous* trees such as oak and ash which lose their leaves in winter, plus coniferous trees such as spruce and pine. And you may see wolves, foxes, deer, and squirrels.

Steppe

This biome is grassland. The climate here is too dry to support forests. Farmers rear cattle, or plough the land to grow crops such as wheat and barley. You'll find small animals like hamsters and mice.

Grass. Again!

Mountain

As you go higher it gets colder, the soil gets thinner, and the vegetation more sparse. Above a certain height – the **tree line** – no trees will grow. You may find mountain goats, deer, lynx, and foxes.

Your turn

1 Explain carefully how these affect climate. (Glossary?)
   a latitude   b altitude   c distance from the sea

2 a Overall, which is the main type of climate in Russia? (**A**?)
   b Which has a milder climate: European Russia, or Siberia?

3 a Page 112 refers to *precipitation*, rather than *rain*. Why?
   b Precipitation in Russia is *generally* low. Give two reasons.

4 Compare the taiga and steppe biomes. In at least five lines!

5 Is there a link between climate and biomes? To decide, compare maps **A** and **B**. Look for matching patterns. Give at least three pieces of evidence to support your answer.

6 Which of those five biomes might be easiest for humans to live in? Write at least five lines, giving your reasons.

This unit is all about the people of Russia – including where they live, and how well off they are.

Population distribution in Russia

Russia has around 146 million people. Map **A** shows how they are spread. The boundary between Europe and Asia is marked in.

Top five cities (millions of people)	
Moscow	12.7
St Petersburg	5.5
Novosibirsk	1.7
Yekaterinburg	1.5
Nizhny Novgorod	1.3

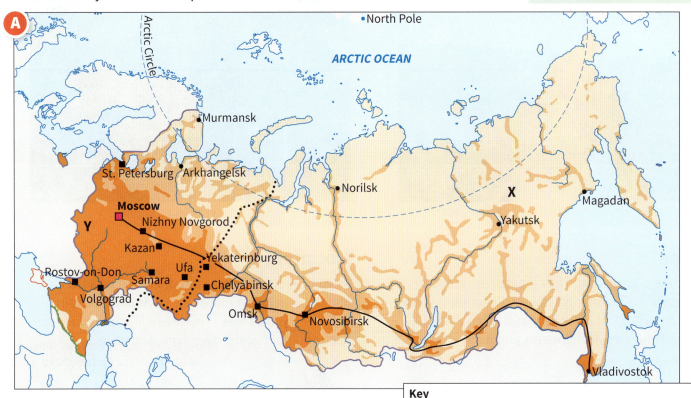

Key

people / sq km		cities			
▢	10–100	▪ capital city		～	river
▢	1–10	■ cities with over 1 million people		—	route of the Trans-Siberian railway
▢	under 1	• selected smaller cities		••••	boundary between Europe and Asia

Note the uneven distribution. In fact about 78 % of the population lives in the European part of Russia. Large areas of Siberia are completely empty.

There are reasons for this uneven distribution, of course. You'll explore them in *Your turn*.

▼ *Moscow, Russia's capital city. The big white skyscraper on the far bank of the River Moskva is an apartment building.*

▼ *St Petersburg, Russia's second largest city. The building with the tall spire is a cathedral. Most of the Tsars are buried within it.*

The people

- As it expanded, the Russian Empire took over peoples with different languages, cultures, and religions. So today, Russia has 185 ethnic groups.
- The Russian ethnic group is the biggest – about 81 % of the population.
- Overall, 75 % of the population is **urban**. (Compare with 84 % for the UK.)
- Russia's **fertility rate** is low, at 1.6 children per woman. So experts predict that the population will fall to 125 million by 2050! The government offers tax breaks and other incentives to couples, to have more children.

What do people do for a living?

Compare these pie charts. What differences do you notice?

B % of workers in each sector

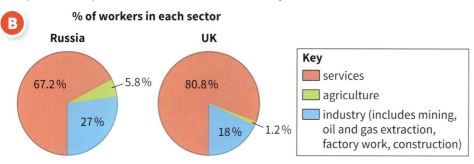

Russia: 67.2 %, 5.8 %, 27 %
UK: 80.8 %, 18 %, 1.2 %

Key
- services
- agriculture
- industry (includes mining, oil and gas extraction, factory work, construction)

▲ *Priests and monks of the Russian Orthodox Church – Russia's main religion. (It is a Christian religion.) Islam is second.*

Rich, or poor?

Look at **C**. GNI per person (PPP) is a measure of how well off people are.

- Russia ranked 50th in the world for GNI per person (PPP) in 2019.
- But the figures are *averages*. In fact, in 2019 …
  - the richest 10 % of Russians held nearly 90 % of Russia's wealth.
  - around 14 % of Russians had only enough to survive.
- So there is great **inequality** in Russia. (There's inequality in every country. But it is more extreme in Russia than in most.)
- Very wealthy Russians are called **oligarchs**. Many bought state-owned businesses cheaply when the USSR broke up, and earn a fortune from them.

C

Country	GNI per person (PPP) in dollars, 2019
Russia	28 270
China	16 760
Germany	59 090
Malawi	1080
UK	49 040
USA	66 080

Your turn

1 **a** In which part of Russia do most people live?
   b Give the population density at: **i** X **ii** Y on map **A**.

2 *Does climate influence the population distribution in Russia?* Compare map **A** in this unit with map **A** on page 112. Look for matching patterns. Then answer the question in italics, giving your evidence.

3 Main rivers are marked on map **A**. Look for patterns that show a link between rivers and population density.
   a Describe what you notice. **b** Then explain it.

4 Map **A** also shows the famous Trans-Siberian railway.
   a Name the cities at the ends of this railway.
   b Repeat question **3**, but this time for the railway.

5 **a** What is the *replacement fertility rate*? (Glossary?)
   b Explain why Russia's population is expected to fall.

6 **B** compares employment in Russia and the UK.
   a Which has a bigger % of workers in farming?
   b Name any four jobs in the services sector. (Glossary?)
   c Russia has a higher % in industry than the UK does. Is it good to have a lot of industry? Explain your logic.

7 This is about the countries in table **C**.
   a Define *GNI per person (PPP)*. (Glossary?)
   b In which country are people wealthiest, on average?
   c Compare the values for Russia, China, and the UK.
   d Is everyone in Russia equally wealthy? Explain.

Here you can find out more about Russia west of the Urals. Just match the numbers on the photos to the numbers on the map.

A reminder

- About a quarter of Russia's land lies west of the Ural Mountains, in Europe.
- This area is where most Russians (78 %) live, and where the largest cities are.
- Overall, the climate is milder than in Siberia.

RUSSIA

Take a tour!

1

Moscow. The capital, with nearly 13 million people. It has factories making cars, helicopters, textiles, and more. The big complex of buildings above is the Kremlin, where the president lives.

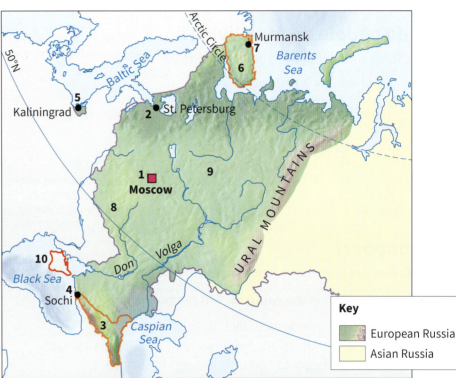

Arctic Circle
50°N
Baltic Sea
Murmansk 7
Barents Sea
6
Kaliningrad 5
St. Petersburg 2
9
1 Moscow
U R A L   M O U N T A I N S
8
Don   Volga
10
Black Sea
4 Sochi
3
Caspian Sea

Key
European Russia
Asian Russia

2

St. Petersburg. Russia's second largest city, with over 5 million people. It has a port, and lots of industry, including shipbuilding. It has many beautiful buildings.

3

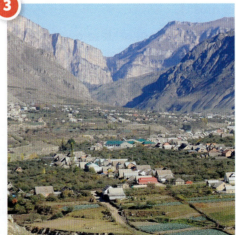

The North Caucasus. Largely a farming region, and largely Muslim. It includes the republic of **Chechnya**. In the past, Chechens have fought in vain for independence from Russia.

4

Sochi. Russia's top holiday resort. On the Black Sea. Warm summers and mild winters. It's also a centre for sports. You can ski in the nearby Caucasus mountains.

5

Kaliningrad. This exclave gives Russia a second port on the Baltic Sea. It is a manufacturing hub for cars, TVs, and other goods. It is smaller than Birmingham in area!

6

The Kola Peninsula. Far north, but quite mild in winter thanks to the **North Atlantic Drift**. It is highly industrial, based on mining for metals. It is heavily polluted too.

7

Murmansk. The Kola Peninsula's main city, and a port. It is ice-free in winter. It is important for fishing. Its population of around 300 000 makes it the biggest city in the Arctic.

8

Farmland. There's lots. Many farms are very large, from when the state owned all the land. Crops include wheat, barley, and potatoes. The best farmland lies south of Moscow.

9

Rural villages. You'll find many villages of wooden houses on your travels. Many are in decline, because people are moving to the cities. Over time, the forest will reclaim the land.

10

Crimea. Under international law, Crimea is part of Ukraine. But Russia controls it. (See page 108.) Sevastopol is a port and naval base. Yalta is a seaside resort.

This is a small selection of what you might see on your travels west of the Urals. You will find much, much more to interest you!

Your turn

1 Look at the photos in this unit. Which *three* places would you most like to visit? Explain your choice.

2 What and where is:
 a Chechnya?   b the Kola Peninsula?
 c Crimea?   d Kaliningrad?

3 Which geographical factors have helped to turn Sochi into Russia's top holiday resort? (Don't forget latitude!)

4 St Petersburg port gets iced over in winter. Murmansk port is further north, in the Arctic – but it is ice-free. Explain why.

5 Look at the rural village in photo **9**. European Russia has many rural villages – but many are in decline. Suggest *three* reasons why people want to move to the cities.

6 Four ports are named in this unit.
 a List them, and say which sea or ocean each port is on.
 b For each port, try to explain its importance to Russia. (Think about where ships can go from there. Page 141?)

7 This unit is about European Russia. What types of landscape would you see in Asian Russia? Describe at least two.

Overall, Russia is warming 2.5 times faster than the global average. How is this affecting Yakutia in Siberia? Read on.

Nightmare on Avtodorozhnaya Street

It's 4 am on 25 June, 2020 in Yakutsk, the capital of Yakutia.

In a small block of flats on the outskirts of the city, everyone is asleep. But suddenly, loud cracking sounds wake them. In panic, they grab some clothes and run outside. They look on in awe. The building has split.

The cause? Not an earthquake. Not an explosion. It's the permafrost!

Like most buildings in Yakutsk, theirs rests on concrete stilts, sunk into the permafrost. This is to keep it stable in summer, when the upper layer thaws. And also to reduce heat transfer from the building to the ground.

But now the building is damaged, like hundreds of others in Yakutia. Because every year, as the climate warms and summers grow longer, the permafrost thaw goes deeper. The nightmare will continue.

▲ An apartment building in Yakutsk, built on stilts. If the permafrost thaws …

More about Yakutia

- **Yakutia** is Russia's biggest administrative region. It is in Siberia. (Its official name is the **Republic of Sakha**.)

- It is huge – about one fifth of Russia, and almost as big as India!

- But it is sparsely populated, with fewer than one million people.

- It is Russia's coldest region. Over 40 % of it lies within the Arctic Circle. It sits on **permafrost**, more than 700 metres deep in places.

- The winters last for eight months, with temperatures below −40 °C. Summer days can reach 26 °C. And now … getting even warmer!

- It is rich in resources: diamonds, gold and other metals, coal, oil, and gas.

- There is farming in the south, along the River Lena and its tributaries. Farmers grow potatoes and other vegetables, and wheat, in the warmer months. Cattle and horses are raised for meat and milk.

▲ Because it's so cold in winter in Yakutia, you can buy milk as frozen discs.

▼ Yakutsk. It lies on the River Lena, which freezes in winter. Population: about 311 000.

▼ Reindeer are raised in northern Yakutia. Their pastures are in the tundra.

▼ A sunny day on a sandy bank of the Lena. Summers are getting hotter.

How is climate change affecting Yakutia?

Overall, Russia is warming faster than the rest of the world. What are the effects on Yakutia?

- Once, in summer, the permafrost would thaw to a depth of 60 cm or so. Now the thaw goes much deeper.

- As it thaws, the soil gets soggy. And often there are big blocks of solid ice in it. When these melt the land sinks permanently, creating marshy hollows and hummocks.

- Water from the thawing permafrost feeds rivers. The Lena is already prone to flooding while its ice melts. Now floods are bigger and more frequent.

- So it is getting harder to grow crops, and graze cattle, in Yakutia. Many farmers have given up, and moved into Yakutsk.

- Wildfires are a problem too. With warmer summers – and more heatwaves – they are becoming more frequent.

- Finally, the thawing permafrost releases tonnes of carbon dioxide and methane. These are greenhouse gases. They speed up climate change. Not only in Yakutia, but across the world.

A major challenge for Russia

It's not just Yakutia. About 65 % of Russia sits on permafrost. Buildings, roads, railways, pipelines and power stations – all are at risk from a deeper thaw. It will cost billions to repair or replace them.

Any benefits?

Many people think that climate change will benefit Russia.

For example, permafrost makes mining and farming difficult. Mining companies say that climate change will make mining easier. Others hope that Russia can grow lots more crops, in the long run.

But meanwhile, those greenhouse gases continue to escape …

▲ F *The deeply thawing permafrost makes pastures lumpy. These native Yakutian horses can cope better than cattle. They are reared for milk as well as meat.*

▲ G *Along river banks, the thawing permafrost exposes the remains of woolly mammoths. They died out in Siberia about 10 000 years ago. Tusks are sold to China.*

▲ H *A petrol tanker falls through the ice on the River Aldan in Yakutia. Frozen rivers are used as roads in winter. But climate change is making this more risky.*

Your turn

1 The average population density in Yakutia is only 0.31 people per sq km. Why is it so low?

2 Even though it is largely empty, Yakutia is of great importance to Russia. Explain why.

3 a Define *permafrost*. (Glossary?)

  b Buildings in Yakutsk are on concrete stilts. Why?

  c Imagine you live in the building in **A**. How might it, and you, be affected, if the permafrost below it thaws?

4 Farming is restricted to the south of Yakutia. Why?

5 Give two ways in which climate change is making life more difficult for farmers in Yakutia today.

6 One result of climate change in Yakutia worries people everywhere. Identify it, and explain why they worry.

7 How might climate change affect this, in Yakutia? Explain.
  a population density   b average wealth per person

8 Choose any two photos in this unit that you can link. For example two on the same theme, or showing a contrast. Explain your choice, and describe what the photos show.

Scientists predict that the Arctic Ocean will become ice-free in summer. So how is Russia responding? Find out here.

The Arctic Ocean

Map **A** shows the **Arctic Ocean** and **Arctic region**. The countries in the Arctic region all have territory within the Arctic Circle. (This passes through a small island off Iceland.)

The Arctic Ocean is iced over in winter. In summer, much of the ice cover melts.

Satellites have been capturing images of the Arctic Ocean for over 40 years. And these confirm a trend: more and more ice is melting in summer. This is linked to **climate change**.

Experts predict that the Arctic Ocean will be ice-free in summer by 2040, or even sooner!

Can the melting ice benefit Russia?

Look again at **A**. Russia has the longest coastline on the Arctic Ocean.

But it is not just coastline. A coastal country has rights over an **exclusive economic zone** or **EEZ**, extending 370 km out from its coast. It can exploit all the resources in this zone. Look at Russia's EEZ in **B**.

So Russia sees a huge opportunity as the summer ice decreases.

- There are vast deposits of oil, gas, and metal ores in the ocean floor.
- There is plenty of fish.
- Tourists will want to visit. They could sail out to the North Pole.
- Goods can be transported on an ice-free ocean between China, Russia, and Europe. Look at the **Northern Sea Route** in **C**.

▲ Polar bears rely on ocean ice to reach their main food: seals. Melting ice is forcing them to scavenge on land.

▼ Russia's EEZ – plus extra it claims as part of its continental shelf (where ocean floor slopes gently from the coast). Note the North Pole!

▼ The Northern Sea Route. It's shorter than using the Suez Canal – for example for shipping from China to the UK.

B Russia's claim to the Arctic Ocean

Russia's EEZ
further claim

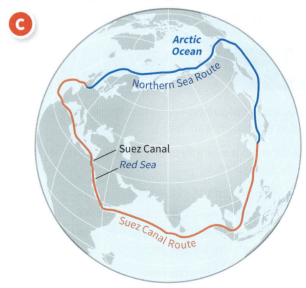

Getting ready to exploit the Arctic Ocean

The Arctic Ocean offers Russia opportunities – and threats. For example, Russia is more open to attack from the ocean when it is ice-free.

So this is how Russia is preparing to exploit the Arctic Ocean.

- It has spent billions on building or improving military bases along the coast, and on islands off the coast. (That makes other countries nervous!)

- It is expanding its fleet of **icebreakers**. These will escort ships along the Northern Sea Route, when there is a risk of ice.

- It has invited foreign companies to help it extract the resources in its EEZ, and develop ports and airports along the coastline.

- It has welcomed investment from China. China now sees the Northern Sea Route as part of its **Belt and Road initiative**.

But note that the ocean will still ice up in winter.

▲ *It's the North Pole! A robot placing a Russian flag at the North Pole in 2007, in support of Russia's Arctic Ocean claims.*

And the dilemma …

Russia plans to extract oil and gas from the Arctic Ocean floor, and sell them.

These fossil fuels are linked closely to climate change.

There is also a risk that extracting materials from the ocean floor will harm ocean **ecosystems**. Food chains in the Arctic Ocean benefit the whole world.

So … it's a dilemma. Is exploiting the Arctic Ocean the way to go? What do you think?

Follow that icebreaker!

▲ *A Russian icebreaker. This one is nuclear-powered. The nuclear fuel lasts for years, so the ship can stay at sea for long periods. Icebreakers force their way through ice, clearing a route for other ships, and helping to keep frozen ports open.*

Your turn

1. Denmark belongs to the Arctic region because of Greenland. (Map **A**.) What is the link between Denmark and Greenland?

2. Only five countries have coastline on the Arctic Ocean. Name them. (Denmark is one …)

3. How is the summer ice cover of the Arctic Ocean changing, and why?

4. a What is an *EEZ*? Include its full name in your answer.
   b Russia claims the largest EEZ in the Arctic region. Why?

5. Study map **B**. Then comment on the size of the total area Russia claims in the Arctic Ocean.

6. a Study **C**. Then explain why China is very interested in an ice-free Northern Sea Route. (Page 141 may help.)
   b Ice usually starts to reform in mid-September. From then, icebreakers are increasingly used. What is an *icebreaker*?

7. Apart from the Northern Sea Route, give two other ways in which Russia can exploit an ice-free Arctic Ocean.

8. Do you agree with this person?
   a First, write two sets of points, one in favour of exploiting the Arctic Ocean, and one against.
   b Then write her a balanced reply.

Ban all claims to the Arctic Ocean now!

6 About Russia

How much have you learned about Russia? Let's see.

check

A

Russia and its surroundings

B

C

1 Map **A** shows Russia and its surroundings.

  a Name the body of water labelled: i 1 ii 2

  b i Name the mountain range labelled 3.

   ii The vast area of Russia to the east of 3 is called … ?

  c i Name the river labelled 4.

   ii Which two biomes does this river pass through?

  d Russia's two largest cities are represented by dots 5 and 6.

   i Name the two cities.

   ii Which dot represents Russia's capital city?

2 **B** shows a bank where permafrost is exposed.

  a Define *permafrost*.

  b About what % of Russia lies on permafrost?

   i about 25 % ii about 65 % iii about 90 %

  c Russia's permafrost is thawing more deeply year by year.

   i Why is this change taking place?

   ii Give one negative impact it has on people in Yakutia.

   iii Give one way it may benefit Yakutia in the future.

   iv Explain why this change in Russia is also affecting *you*.

3 Look again at map **A**. Two of Russia's regions are marked in. The table below compares population densities.

Region or country	Average number of people per sq km
Central region, Russia	61
Yakutia region, Russia	0.3
Russia overall	9
UK	275

  a Give the main reason why the population density is so low in Yakutia, compared with Central region.

  b Climate change may lead to a rise in population in Yakutia, including along its coastline. Suggest two reasons.

  c About how many times greater is population density in the UK than in Russia? Calculate.

  d Suggest one way in which its enormous size may make Russia difficult to govern.

4 **C** shows a Russian holiday resort on the Black Sea. Lots of families come here. Many have only one child.

  a The fertility rate in Russia is around 1.6. This is likely to lead to a fall in population. Explain why.

  b Suggest two problems that a fall in population might cause, for Russia.

  c There is great inequality in Russia. Many people can't afford to go on holiday. Many live in poverty.

   i Define *inequality*.

   ii Inequality in Russia increased greatly after 1991. Suggest a reason.

   iii Suggest one harmful impact of inequality in a country.

5 a Using pie chart **C** on page 109, give evidence that Russia relies heavily on selling fossil fuels to other countries.

  b In future, Russia may be unable to rely on selling fossil fuels. Give a reason.

6 How might Russia benefit from climate change? And do the benefits outweigh any damage that is done:

  a for Russia? b for the whole world?

Write *at least* half a page in your answer. There's a lot to say!

7 The Middle East

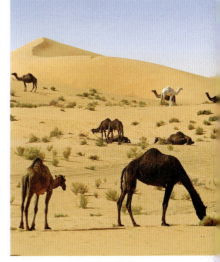

The Middle East is often in the news. Where and what is it? Find out in this unit.

Where is the Middle East?

It is an important region in a strategic location – where Asia, Africa, and Europe meet! Look at **A**.

Note how it is shaped like a tilted X, with one leg in Africa. And note how it is largely surrounded by water.

Middle East is an odd name for this region, since most of it is in south west Asia. But the name came into use over 100 years ago, and it stuck.

The countries of the Middle East

The Middle East consists of 16 countries, and the State of Palestine. Look at **B**. Have you heard of them all?

A

Key
outline of the Middle East region

Arctic Circle
EUROPE
ASIA
AFRICA
Tropic of Cancer
Equator

B

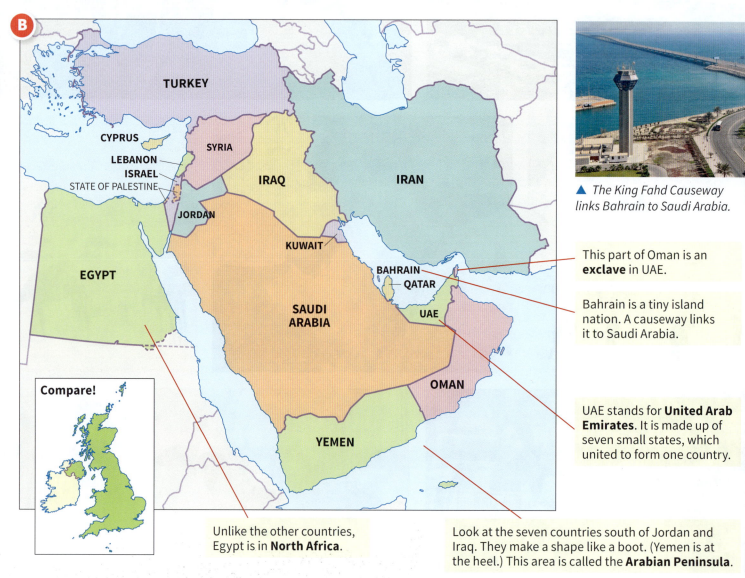

▲ *The King Fahd Causeway links Bahrain to Saudi Arabia.*

This part of Oman is an **exclave** in UAE.

Bahrain is a tiny island nation. A causeway links it to Saudi Arabia.

UAE stands for **United Arab Emirates**. It is made up of seven small states, which united to form one country.

Compare!

Unlike the other countries, Egypt is in **North Africa**.

Look at the seven countries south of Jordan and Iraq. They make a shape like a boot. (Yemen is at the heel.) This area is called the **Arabian Peninsula**.

TURKEY
CYPRUS
SYRIA
LEBANON
ISRAEL
STATE OF PALESTINE
JORDAN
IRAQ
IRAN
KUWAIT
BAHRAIN
QATAR
EGYPT
SAUDI ARABIA
UAE
OMAN
YEMEN

Key points about the Middle East

- Most of it is desert, or dry lands.

- It has very large reserves of oil and gas: about 48% of Earth's known oil reserves, and 38% of its gas reserves! This has helped to make it an important region.

- But some countries have much more oil and gas than others. And some – like Lebanon and Cyprus – are only beginning to exploit their reserves.

- It has some of the world's wealthiest countries, thanks to selling oil and gas.

- Sadly, the region also has a great deal of conflict.

- There is much poverty too, mostly linked to conflict.

▲ *Fertile farmland in Oman. Agriculture depends on irrigation.*

But there is more to it …

There is more to the Middle East than desert, and oil, and gas, and conflict!

- You will find fertile farmland, sunny beaches, mountains where you can ski in winter, and even some glaciers.

- You will find a rich mix of cultures and ethnic groups, with different foods, music, and traditions. Many are linked to the ancient trade between China, India, and Europe, along the Great Silk Road.

- Three world religions began in this region: Judaism, Christianity, and Islam.

- Most people in the region are Muslim. In Israel, around 75% of the population are Jews. In Lebanon, around 41% are Christians. There are small communities of Jews and Christians elsewhere.

▲ *Dubai in the UAE: a city built in the desert.*

Links with Britain

Much of the Middle East was once part of the Islamic Ottoman Empire.

In 1914, the Ottoman Empire entered World War I, fighting against the **Allies** (Britain, France, Russia, and others). The Allies won.

After the war, Britain and France carved up the remains of the empire between them, into areas to control or influence. **C** shows areas linked to Britain. This led to much unrest in those areas. Eventually, they became fully independent.

C

| under British protection/ control by 1922 |

▲ *British involvement in the Middle East.*

Your turn

1 Here are the names of some Middle East countries. You have to unjumble them!

  a EYTKRU   **b** AYRIS   **c** NIAR   **d** ARQAT

  e LNBEONA   **f** PEYGT   **g** MEENY   **h** MOAN

2 Look at the Middle East countries, on map **B**.

  a Which is the biggest?   **b** Which is second biggest?

  c Which is the smallest?   **d** Which two are islands?

  e Which border Iraq?   **f** Which border Syria?

3 What is the *Arabian Peninsula*?

4 Straight lines were drawn on a map to create some country borders in the Middle East. Which ones? (Map **B**.)

5 Name *four* Middle East countries which Britain had some control or influence over, by 1922. (Maps **C** and **B**.)

6 You are a geography teacher. List five key points about the Middle East that you would want your students to remember.

Deserts, mountain ranges, seas, famous rivers, volcanoes. The Middle East has them all. Find out more here.

The physical features of the Middle East

See how much you can discover about the Middle East from this map.

A

Black Sea
PONTIC MOUNTAINS
TURKEY
Mt. Ararat 5133 m
Caspian Sea
TAURUS MOUNTAINS
CYPRUS
Euphrates
Tigris
ELBURZ MTS.
Mt. Damavand 5671 m
SYRIA
LEBANON
ISRAEL
STATE OF PALESTINE
Mediterranean Sea
IRAQ
IRAN
ZAGROS MOUNTAINS
Suez Canal
JORDAN
KUWAIT
Arabian Gulf
Nile
Western Desert
EGYPT
HEJAZ MOUNTAINS
Arabian Desert
BAHRAIN
QATAR
Strait of Hormuz
Gulf of Oman
Tropic of Cancer
UAE
SAUDI ARABIA
OMAN
Red Sea
Rub' Al Khali
YEMEN
Arabian Sea
Gulf of Aden
Bab-el-Mandeb Strait

0   500 km

Key

relief
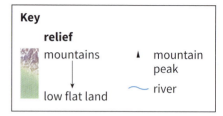
mountains
▲ mountain peak
low flat land
~ river

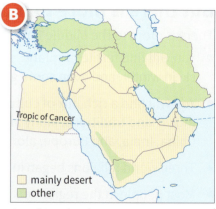

B

Tropic of Cancer

☐ mainly desert
☐ other

▲ Where the desert areas are.

▲ *Mount Damavand in Iran – the highest mountain in the Middle East – is a volcano. It last erupted over 7000 years ago. It puffs steam and gas.*

Mountainous areas

Look at the pattern of mountains in Turkey, Iran, and the Arabian Peninsula.

- **Mount Ararat** is Turkey's tallest mountain. According to legend, it is where Noah's Ark landed, at the end of the great world flood.

- **Mount Damavand** in Iran is the Middle East's highest peak (5671 m).

Some world-famous rivers

- The **Nile** is the world's longest river (6650 km). Egypt is the final country on its long journey to the Mediterranean.

- Now find the **Euphrates** and **Tigris** rivers. They rise in Turkey and flow south. They join to form the **Shatt Al Arab** before reaching the Arabian Gulf. In ancient times, the fertile region between them was called **Mesopotamia**.

Deserts

Much of the Middle East is desert. Compare maps **A** and **B**.

- Over 90% of Egypt is desert – part of the Sahara Desert, which stretches all the way across North Africa from the Atlantic Ocean to the Red Sea.

- The **Arabian Desert** covers most of the Arabian Peninsula. The **Rub' Al Khali**, or Empty Quarter, is its largest continuous stretch of sand.

Surrounded by seas

- The **Caspian Sea**, which borders Iran, is like a huge lake, but salty.

- The **Black Sea** flows through narrow straits into the **Mediterranean Sea**.

- Look at the **Red Sea**. At the northern end is Egypt's **Suez Canal**, leading to the Mediterranean. The other end opens into the **Gulf of Aden**.

- The **Arabian Gulf** opens to the **Gulf of Oman** at the **Strait of Hormuz**.

- The Gulf of Oman and Gulf of Aden lead to the **Arabian Sea**.

Choke points!

90 % of world trade is by sea.

There are several **maritime choke points** around the world, where ships pass through quite narrow passages. These three are in the Middle East: the **Suez Canal**, the **Bab el-Mandeb Strait**, and the **Strait of Hormuz**. Find them on **A**. Sea trade between Asia and Europe depends on them.

Choke points are vulnerable. For example, enemies could block them.

Tectonic activity

Several tectonic plates meet around the Middle East. See the map on page 91. So that means earthquakes and volcanoes. In particular, Turkey and Iran experience many earthquakes.

Table **C** shows the numbers of volcanoes that have erupted in the last 10 000 years. They are now dormant – but not extinct. Some could erupt again.

Note that the Arabian plate is moving away from the African plate, under the Red Sea. That explains why there are volcanoes in the Red Sea – and why it is getting wider!

▲ The Suez Canal: 193 km long, 205 m wide. Ships pay Egypt to use it – about 50 ships a day, carrying cargo worth up to $9 billion.

▲ There are many islands in the Red Sea. This eruption in 2012 created a new one! It belongs to Yemen.

C

Country	Number of volcanoes
Yemen	11
Turkey	10
Saudi Arabia	9
Iran	6
Syria	2

▲ Volcanoes in the Middle East that have erupted in the last 10 000 years.

Your turn

1 a Name *three* Middle Eastern countries where much of the land is mountainous.

  b In which country will you find: **i** the Zagros Mountains?
     ii the Hejaz Mountains?   **iii** the Taurus Mountains?

  c What and where is the highest mountain in the region?

2 a List the bodies of water which border the Middle East region.

  b Which two from **a** are linked by the Strait of Hormuz?

3 The Arabian Sea is part of an ocean. Which one? (Page 140?)

4 a Name the three main rivers of the Middle East.

  b Saudi Arabia has no permanent rivers. Suggest a reason.

5 Write a paragraph on the Suez Canal. Include its location!

6 What and where is the *Rub' Al Khali*?

7 a The Red Sea is slowly getting wider. Why?

  b Turkey is prone to earthquakes. Why?

  c Which Middle East country has had most volcanic eruptions in the last 10 000 years?

Climate, plants, and animals are closely linked. Find out here about the climate zones and biomes in the Middle East.

The climate of the Middle East

The Middle East is hot and dry overall. Let's see why.

Temperature

- The Tropic of Cancer runs through the Middle East, as map **B** shows. That's why the region is hot. It is not so far from the Equator.

- But as you go up the mountains, it gets cooler. You will even find small glaciers in the mountains of Turkey and Iran.

Rainfall

And now … a mystery. The Middle East is largely surrounded by water – but it gets so little rain that most of it is desert. Why? Look at diagram **A**.

- The land at the Equator gets hot, and heats the air. The warm moist air rises. It cools, and its water vapour condenses. Heavy rain falls at the Equator.

- As the warm air rises, colder air flows towards the Equator to take its place.

- The risen air, which is now dry, gets pushed out of the way. It descends again as cool dry air around 30° north and south of the Equator. And cool dry air descending means … little or no rain.

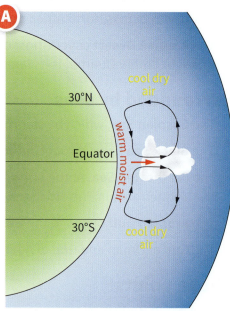

▲ How the air circulates between the Equator and 30° N and S. (This is the Hadley circulation cell that you met in geog.2.)

A climate map for the Middle East

Map **B** shows the three major climate zones of the Middle East.

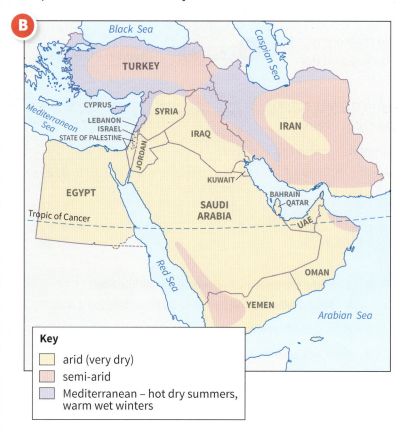

Key
- arid (very dry)
- semi-arid
- Mediterranean – hot dry summers, warm wet winters

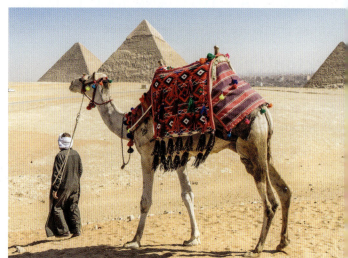

▲ The pyramids near Cairo in Egypt. In an arid landscape, thanks to dry descending air. (Cairo is at latitude 30 °N.)

Why …
… do camels have long eyelashes?

Did you know?
- Egypt has over 100 pyramids.
- Most were built as tombs for the Pharaohs (the rulers of Ancient Egypt).

The biomes of the Middle East

A **biome** is an area that shares similar natural vegetation, and animals. Biomes are closely linked to climate zones. So here are the main biomes of the Middle East:

In desert areas …

Vegetation is sparse. In some areas there is none. Plants have tough spiky leaves to conserve moisture. You might see Arabian oryx (below). And wild camels, sand cats, snakes, scorpions, eagles, and more.

In semi-arid areas …

These have grass, and some low bushes, but few trees. (They are called **steppe**.) People herd sheep and goats. Wild animals include wolves, sand foxes, wild cats, gazelles, and wild boar.

In Mediterranean areas …

This forest of cypresses is in Turkey, by the Mediterranean Sea. You'll also find deciduous forests, and wild shrubs and flowers. You might spot bears, hyena, deer, squirrels, hamsters …

▲ *You will also find wonderful ecosystems in the seas around the Middle East. This is a coral reef in the Red Sea.*

Your turn

1 Write out this paragraph about the Middle East. Fill in the blanks using terms from the white box.

   The Middle East is _____ overall, because of its _____ . It is also dry, because it lies where dry air _____ , after losing _____ over the _____ . Because most places have little rain, vegetation tends to be _____ . But some places have enough rain for _____ to grow.

sparse	wind	Equator	longitude	forests	wet
latitude	hot	cold	ascends	descends	moisture

2 a Where in the Middle East will you find warm wet winters?

   b Overall, the Middle East has a shortage of fresh water. Why?

3 Explain why:

   a there are no natural forests in Oman

   b Saudia Arabia has to import most of the food it needs

4 You are going to the Middle East on holiday. Which climate zone will you choose? Explain why, in at least 5 lines.

How are people spread around the Middle East? And how do they compare for wealth? Find answers to these questions, and more.

Population and population density

Table **A** shows the populations of the Middle East countries. How do they compare with the UK, and London?

Map **B** shows population densities, and capital cities. As you'd expect, hardly anyone lives in the desert areas. But what do you notice about Egypt?

A

Populations in the Middle East (millions), 2020	
Egypt	102.5
Iran	90.0
Turkey	84.3
Iraq	40.2
Saudi Arabia	34.8
Yemen	30.3
Syria	17.5
Jordan	10.2
UAE	9.9
Israel	9.3
Lebanon	6.8
State of Palestine	5.1
Oman	5.1
Kuwait	4.3
Qatar	2.9
Bahrain	1.7
Cyprus	1.2
Total	**456.1 million**
UK	67.9
London	9.3

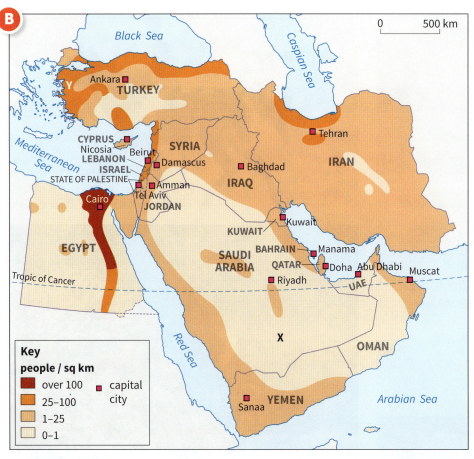

The people

The Middle East has many different ethnic groups. Some are across several countries. Map **C** shows only the larger groups.

- **Arabs** are by far the biggest group. They speak Arabic. Look at the countries which are mainly Arab.

- 80% of the people in Turkey are ethnic **Turks**, and speak Turkish. Most of the rest are **Kurds**, who speak Kurdish.

- In Iran, over 60% of the people are **Persian**. Iran has many other ethnic groups too. Most people speak Persian.

- Kurds are found in Turkey, Iraq, Iran, and Syria. Some Kurds want all the neighbouring Kurdish regions to unite as an independent country.

The economies: exports

Exports vary from country to country, as you would expect.

- Exports of oil and gas dominate in the region. Some countries also sell related products – such as plastics (based on oil) and fertilisers (which use gas as a raw material for making the key chemical, ammonia).

- There are other exports too. For example, cars from Turkey and Saudi Arabia, and clothing and textiles from Turkey, Jordan, and Egypt. Israel exports high-tech equipment, and pharmaceuticals.

- Agricultural exports are important for those countries with more fertile land, such as Turkey, and Egypt (along the Nile).

E Top 10 in the Middle East for reserves

	oil	gas
1	Saudi Arabia	Iran
2	Iran	Qatar
3	Iraq	Saudi Arabia
4	Kuwait	UAE
5	UAE	Iraq
6	Qatar	Egypt
7	Oman	Kuwait
8	Yemen	Oman
9	Syria	Yemen
10	Bahrain	Syria

▲ The top 10 for known reserves. (But this does not tell you how much they export!)

The economies: wealth

Check out chart **D**, for a recent year. In which country are people wealthiest, on average? How do they compare with people in the UK?

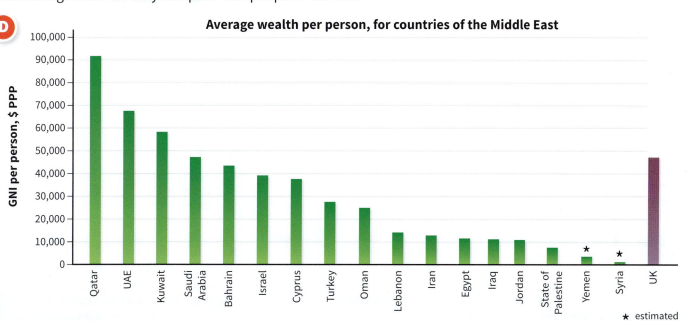

D Average wealth per person, for countries of the Middle East

★ estimated

Your turn

1. Make a table with two columns, labelled *Country* and *Capital city*. Then fill it for *at least* twelve Middle East countries.

2. a For map **B**, *describe* the pattern of population density in Egypt, and then *explain* it. (Map **A** on page 126 may help.)

   b *People in the Middle East avoid living in the mountains.* True or false? Give evidence – and suggest a reason.

   c With the help of earlier maps, explain why the population density is higher at **Y** than at **X**, on map **B**.

3. Table **A** shows the populations of the Middle East countries.

   a Name the three most *populous* countries. (Glossary?)

   b Name three Middle East countries:
      i with larger populations than the UK
      ii with between 4 and 9 million people
      iii with fewer people than London

4. a From **C**, list the four largest ethnic groups by *area*, in order, largest area first. Give the language for each group.

   b Name five countries where most people are Arab.

5. Are oil and gas the only exports of the Middle East? Explain.

6. **D** shows how wealth per person varies across the Middle East.

   a i In which country are people wealthiest?
      ii What is the population of that country?

   b i What can you say about people in Syria, from **D**?
      ii What is the population of Syria?

   c Suggest a reason for the low value for Syria. (Page 135?)

   d In which countries are people wealthier than in the UK?

7. *There is great inequality in wealth across the Middle East.* Discuss, giving evidence from **D**.

The Arabian Peninsula faces some special challenges. Find out more here.

The Arabian Peninsula

The Arabian Peninsula is fascinating, for many reasons.

- It has seven countries, with a total poulation of around 89 million. (The UK has around 68 million.) Most of the population – over 73 % – is in Saudi Arabia and Yemen.

- The peninsula is mostly desert.

- All its countries, except Yemen, are monarchies. They are ruled by Kings or Emirs or Sultans. In Saudia Arabi and Oman, the rulers are **absolute monarchs**, with full control over the country.

- Its countries have developed rapidly since oil was discovered (first in Saudi Arabia, in 1938). The money from oil and gas has improved people's lives.

- But Yemen is an exception. Although it has some oil and gas, it is one of the world's poorest countries. It has suffered years of conflict.

▲ *Before oil and gas were found, the people of the Arabian Peninsula were mostly nomads. They moved with their sheep and goats and camels. Now they are settled.*

What are the challenges?

The Arabian Peninsula faces a storm of challenges!

① **Water stress.** The main source of fresh water is a huge aquifer system under the peninsula. It is being depleted.

② **Food risk.** Less than 2 % of the land is naturally arable. Most food has to be imported.

③ **A fast-growing population.** Over 1 million more people a year!

④ **Climate change.** The peninsula is getting hotter. The north will get even less rain than now, but rainfall may increase in the south.

⑤ **A fall in demand for oil and gas.** The world is turning from oil and gas because of climate change. But these countries depend on selling them!

Let's see how some of these challenges are being tackled.

Tackling water stress

- Fresh drinking water is widely obtained by the **desalination** of sea water. (But even with cheap oil and gas as fuel, the process is expensive.)

- Waste water is widely cleaned up, and used for crops.

- The UAE carries out **cloud seeding**. Planes spray tiny salt crystals into clouds, to encourage water droplets to grow larger, and fall as rain.

- People do not have to pay much for the water they use. So they waste it! Advisers say that water charges must be increased, to prevent waste.

▶ *Water: more precious than oil? Fountains at the Burj Khalifa, the world's tallest building, in Dubai.*

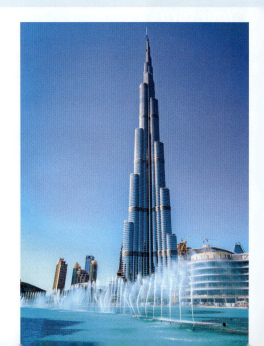

Tackling the food issue

Food is a big worry for these countries. What if they can't obtain enough food in the future?

- So in Saudi Arabia, the government gave out grants to companies to turn desert into farmland. This meant heavily use of fertilisers, and water from the aquifer.

- But the farms are very expensive to run. It is cheaper to import the food than grow it!

- So Saudi Arabia has also invested in farmland in Australia, in Sudan in Africa, and perhaps in other countries. The idea is to import the crops from it.

Can hydroponics help?

Several peninsula countries are exploring **hydroponics**. Plants are grown in greenhouses, in water containing the nutrients they need. (Page 33.)

- The greenhouses are designed to be quite cool inside, and humid.

- The water can be obtained by desalinating seawater, using solar power.

- The water and unused nutrients can be recycled.

- There is plenty of empty desert for greenhouses, and lots of sunshine!

Hydroponics could play a big part in the food supply in the future.

Is there life after oil and gas?

For decades, governments of the peninsula have been saving up money from oil and gas sales, as security for the future.

- They use some to buy shares in companies in other countries.

- They use more to **diversify** their own economies. For example:
  - Dubai in the UAE is now a financial hub for the Middle East, and a world shopping destination
  - Qatar plans to be a hub for culture and sports
  - Saudi Arabia is promoting tourism.

- They want to develop high-tech industries.

- They want more people to start their own businesses. This goal may take time, because people are paid well as employees.

▲ *This satellite image shows farming in the desert in Saudi Arabia. The green circles show irrigated crops, as on page 28. But the water level in the aquifer below them is falling by up to 6 metres a year.*

▲ *Cleaning windows at the Khalifa International Stadium in Qatar, the main venue for the 2022 World Cup.*

Your turn

1   List the countries of the Arabian Peninsula.

2   a   What is an *absolute monarch*? (Glossary?)
    b   Do we have an absolute monarch in the UK? Explain.

3   Think about those five challenges facing the Arabian Peninsula. Pick out what you think are the two main challenges. Explain your choice.

4   A large aquifer system underlies the Arabian Peninsula. Explain why its water levels are falling. Give *two* reasons.

5   a   Why has Saudi Arabia invested in land in Sudan?
    b   Some people in Sudan are not happy about this. Suggest a reason.

6   Is hydroponics a more *sustainable* way to provide food than planting crops in the desert? Explain your answer.

7   Explain the term *diversify*, and give an example.

8   A shortage of water could ruin all plans to diversify, in the Arabian Peninsula. Do you agree? Explain.

Some countries of the Middle East have been torn apart by conflict. Find out more here.

An unstable region

The Middle East has long been – and still is – an unstable region. Why? Here are some reasons:

- Borders created by Britain and France caused some problems. (Page 125.) Different ethnic groups were forced together – or split up, like the Kurds.

- Most people in the Middle East are Muslim. But Islam has several branches. The two main ones are **Sunni** and **Shia**. Look at map **A**. Tensions between Sunni and Shia Muslims have led to violence.

- Many rebel groups have operated in the Middle East, generating conflict. Some are **Islamic extremists**. Two examples are **Islamic State** (or IS), and **Al-Qaeda.** Most Muslims say these do not represent Islam.

- Foreign countries continually involve themselves in the Middle East. Why? Reasons include:
  - to make sure they continue to get oil and gas
  - to prevent extremism from spreading
  - to sell things; for example, many countries, including the UK, earn a great deal by selling weapons to the Middle East
  - to gain influence; for example, the USA, China, and Russia compete for influence around the world.

  Factors like these shape the relationships between countries. This is called **geopolitics**.

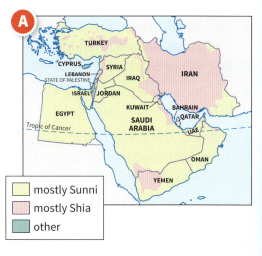

A

mostly Sunni
mostly Shia
other

The conflict hotspots

Page 135 shows the main conflicts. *Several countries have been involved in most of them*, each for their own reasons. So things get very complex.

For example, in the civil war in Syria, Russia and Iran helped the government. The USA, UK, France and Turkey helped rebel groups. 'Help' included sending in troops, carrying out air raids, and providing weapons, money, and training.

When a country is torn apart by conflict, its development is reversed. It is the ordinary people who suffer most.

▲ *Refugees from Syria, living in Turkey. Families fled here to save their lives.*

Your turn

1 Look at map **A**. Which branch of Islam is the main one in:
  a Saudi Arabia? b Iran? c Iraq? d Oman?

2 Define these terms. (Glossary?)
  a *extremists* b *geopolitics* c *sanctions*

3 a Outline four reasons why foreign countries get involved in the Middle East.
  b Why do countries want influence in other countries?
  c Which of the reasons in **a** are connected to trade?

4 Why do Lebanon and Turkey have so many refugees? (Map!)

5 How might the civil war have affected young people in Yemen? Answer in text, or as a spider map, or …

6 a Compare **D** and **E** on page 131. Give evidence that:
    i Iran's economy has been affected by sanctions
    ii war has had an impact on Iraq's economy
  b Two values in **D** were *estimates*. Suggest a reason, using information from page 135.

Turkey

- ongoing conflict between the government and Kurdish rebels who want self-rule (about one-fifth of Turkey's population are Kurds)
- took in more refugees than any other country in the world: 4 million, most from Syria

Iran

- suspected of helping extremists, and developing nuclear weapons
- other countries use sanctions to punish it; these limit oil exports, and other activities
- some sanctions still in place in 2021
- Iran's economy hit hard by sanctions

Syria

- civil war 2011 – 2020? (ceasefire in 2020)
- around 400 000 Syrians killed; over 5 million fled to other countries (mainly Turkey and Lebanon) as refugees

Lebanon

- civil war 1975 – 1991
- other conflicts since then
- fragile, economy struggling
- took in over 1.7 million refugees, most from Syria

Israel and the State of Palestine

- land historically known as **Palestine**
- Jewish people offered a homeland there by Britain, after World War I; it became **Israel**
- result was conflict with the Arabs who were already living in Palestine; it continues today
- Arabs live in two separate areas, **Gaza** and the **West Bank**; these form the **State of Palestine**
- settlements built by Israel in the West Bank (illegal under international law) cause much conflict

Iraq

- invaded in 2003 by forces led by the USA and UK, who thought its leader, Saddam Hussein, was a danger to the world
- by the end of the war in 2011, IS had sprung up
- further war to defeat IS, 2013 – 2017
- Iraq still fragile today

Yemen

- a civil war began in 2014
- mainly Houthi rebels fighting government forces
- but several groups involved
- by 2021, over 100 000 Yemenis killed, and over 85 000 dead from famine caused by the war

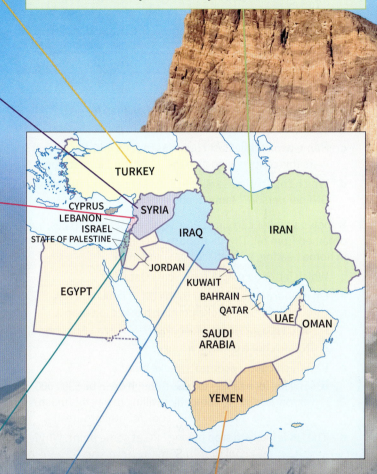

TURKEY
CYPRUS
LEBANON
ISRAEL
STATE OF PALESTINE
SYRIA
IRAQ
IRAN
JORDAN
EGYPT
KUWAIT
BAHRAIN
QATAR
UAE
OMAN
SAUDI ARABIA
YEMEN

7 **The Middle East** *check* ✓

How much have you learned about the Middle East? Let's see.

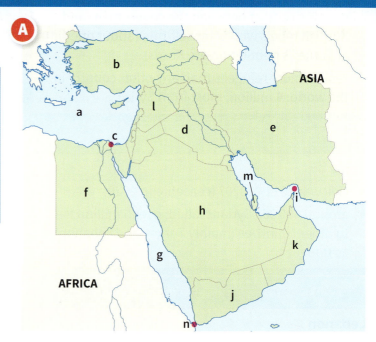

A

1 Map **A** shows countries of the Middle East.

a Copy and complete this table. One label has been filled in.

country	label on map A	capital city
Egypt		
Saudi Arabia	*h*	
Turkey		
Yemen		
Iraq		
Iran		
Oman		

b i Name the seas marked **g** and **a** on map **A**.

   ii Explain why there are volcanoes in **g**.

c Name two Middle East countries that have suffered civil wars this century, and give the letter used for each on **A**.

d Which country in the Middle East is wealthiest, in terms of GNI per person? Name it, and give its letter on **A**.

2 **B** shows a tanker carrying LPG – liquified petroleum gas, obtained from oil. The UK imports LPG from Qatar, for use in home heating, as barbecue gas, and more.

a Which letter on map **A** represents Qatar?

b On its journey from Qatar to the UK, a tanker passes through maritime choke points **i**, **n** and **c**, on map **A**.

   i What is a *maritime choke point*?

   ii Name choke points **i**, **n** and **c**.

c i Which country controls choke point **c**?

   ii The fee for a tanker to pass through **c** can be $300 000 or more. Why are companies willing to pay this much? (Pages 140–141 may help. Look for other routes!)

d The UK is likely to stop buying LPG at some point. Why?

B

3 a Table **C** shows the two most populous countries of the Arabian Peninsula, with data for 2020.

   i Which country is more densely populated, overall? Explain your answer, with the help of map **A**.

   ii About how many times wealthier are people in Saudi Arabia than in Yemen? Choose:
   7 times   24 times   47 times

   iii Suggest *two* reasons for this difference in wealth.

4 Saudi Arabia has on average only 70 mm of rainfall a year. But farmers do grow some crops.

a Where do farmers get fresh water to irrigate these crops?

b The source in **a** may not always be available. Why not?

c Many Middle Eastern countries obtain fresh water from sea water. What is this process called?

d Another Middle Eastern country has about 400 mm of precipitation a year. It is either Oman or Turkey. Which one? Explain your choice.

C

Country	Population (millions)	Average wealth per person (GNI per person $ PPP)
Yemen	30.3	980 (estimated)
Saudi Arabia	34.8	46 270

5 In what ways have the resources of oil and gas:

a benefited the Middle East?

b helped, probably, to make the region more unstable?

6 a World demand for oil and gas will decline. Why?

b This decline is a special challenge for the desert countries of the Arabian Peninsula. Explain why.

7 *Studying the Middle East has helped me to understand how the world is interconnected.*
To what extent do you agree with this statement? Give examples to support your answer.

Command words: a summary

All through *geog.123* there are questions with *command words* that tell you how to answer them. Later, you'll meet these words in exams too. So it's good to get used to them.

Here's a summary of the command words to help you, in alphabetical order. The command words and their definitions are in red. (And note that we cover them in more detail in Chapter 1 of *geog.1*.)

Assess
Weigh everything up and make a judgement.
For example, assess the impact of an earthquake on a city. You must always include the evidence you based your judgement on.

Calculate
Do some maths, to get the answer!
Always give the unit in your answer. For example, *5 km* or *11 people* or *15 days*. (Not just *5* or *11* or *15*.)

Compare
Say what is the same, and different, about two things.
For example, say which one is bigger. Always mention both things in your answer.

Copy and complete
Copy this, filling in all the blanks.
Fill in using the words and terms that make sense!

Define
Write down the meaning.
Keep your answer clear and simple.

Describe
Write a description.
For example, describe what you see, or the steps in a process. You do not need to give reasons for anything.

Discuss
Look at something from different angles, and give key points about it.
For example, you could give its good and bad points, or its benefits and drawbacks.

Draw
Like it says – draw!
For example, draw a diagram, or sketch map, or bar chart, or line graph. Use a ruler for straight lines. Be accurate with bar charts and graphs. And try to be quick!

Evaluate
Judge how successful or worthwhile something is.
You should say what has been good and bad about it, and give your final opinion.

Examine
Look at each part and say how it contributes.
For example, examine how different processes work together to form an oxbow lake.

Explain
Make something clear and easy to understand.
For example, explain how a meander forms.

Give
Come up with an answer, from what you've learned.
Keep it clear and simple.

Identify
Pick out the thing, and give its name.
For example, identify a landform on a map.

Justify
Give reasons to support the choice or decision you made.
For example, give reasons why you agree with a statement.

Label
Add labels!
For example, label a diagram. The aim is to make it clear and easy to understand. So keep labels short and simple!

Name
Write the name of the thing you are asked about.
Easy! You do not need to write a full sentence.

Outline
Set out the main points.
Stick to the main points. You don't need to give details.

State
Give the answer in clear terms.
State is often used in place of *Give* or *Identify* or even *Calculate or Count*. Be sure to answer clearly.

Suggest
Come up with a possible reason or plan.
Use your common sense!

To what extent .... ?
How much does it contribute, or how important / true is it?
Make a judgement, and give the evidence you based it on.

Ordnance Survey symbols

ROADS AND PATHS   1:25 000

M 1 or A 6(M)	Motorway
A 35	Dual carriageway
A 30	Main road
B 3074	Secondary road
	Narrow road with passing places
	Road under construction
	Road generally more than 4 m wide
	Road generally less than 4 m wide
	Other road, drive or track, fenced and unfenced
	Gradient: steeper than 1 in 5; 1 in 7 to 1 in 5
Ferry	Ferry; Ferry P – passenger only
	Path

PUBLIC RIGHTS OF WAY

1:25 000	1:50 000	
		Footpath
		Bridleway
+ + + + +		Byway open to all traffic
		Restricted bridleway

RAILWAYS   1:25 000

	Multiple track
	Single track
	Narrow gauge/Light rapid transit system
	Road over; road under; level crossing
	Cutting; tunnel; embankment
	Station, open to passengers; siding

BOUNDARIES   1:50 000

	National
	District
	County, Unitary Authority, Metropolitan District or London Borough
	National Park

HEIGHTS/ROCK FEATURES   1:50 000

50	Contour lines
· 144	Spot height to the nearest metre above sea level

outcrop   cliff   scree

ABBREVIATIONS   1:25 000 and 1:50 000

PO / P	Post office	PC	Public convenience (rural areas)
PH	Public house	TH	Town Hall, Guildhall or equivalent
MS	Milestone	Sch	School
MP	Milepost	Coll	College
CH	Clubhouse	Mus	Museum
CG	Cattlegrid	Cemy	Cemetery
Fm	Farm	Hosp	Hospital

ANTIQUITIES   1:25 000 and 1:50 000

VILLA	Roman		Battlefield (with date)
Castle	Non-Roman		Visible earthwork

LAND FEATURES   1:25 000 and/or 1:50 000

ruin	Buildings
	Public building
	Bus or coach station
	Place of Worship (current or former) — with tower; with spire, minaret or dome; without such additions
	Chimney or tower
	Glass structure
H	Heliport
△	Triangulation pillar
	Mast
	Wind pump / wind turbine
	Windmill
+	Graticule intersection
	Cutting, embankment
	Quarry
	Spoil heap, refuse tip or dump
	Coniferous wood
	Non-coniferous wood
	Mixed wood
	Orchard
	Park or ornamental ground
	Forestry Commission access land
	National Trust – always open
	National Trust, limited access, observe local signs
	National Trust for Scotland

WATER FEATURES   1:25 000 and/or 1:50 000

Marsh or salting   Towpath   Lock   Cliff   Slopes   High water mark
Aqueduct   Canal   Ford   Flat rock   Low water mark
Weir   Normal tidal limit   Sand   Lighthouse (in use)
Lake   Bridge   Dunes   Lighthouse (disused)   Beacon
Footbridge   Mud   Shingle
Canal (dry)

TOURIST INFORMATION   1:25 000 and/or 1:50 000

P	Parking
V	Visitor centre
i	Information centre
	Recreation/leisure/sports centre
	Telephone
	Camp site/Caravan site
	Golf course or links
	Viewpoint
PC	Public convenience (toilet)
	Picnic site
	Pub/s
	Cathedral/Abbey
M	Museum
	Castle/fort
	Building of historic interest
	English Heritage
	Garden
	Nature reserve
	Water activities
	Fishing
☆	Other tourist feature

© Crown copyright

Map of the British Isles

Key

- `- - - -` international boundary
- `———` national boundary
- river
- lake
- ▲ highest point in the UK

towns
- ■ largest cities
- ● large cities and towns

Land height
measured in metres above sea level

- more than 1000 m
- 500 - 1000 m
- 200 - 500 m
- 100 - 200 m
- less than 100 m
- land below sea level

● a label this colour shows a place you study in this book

Scale
1: 4 500 000

One centimetre on the map represents 45 kilometres on the ground.

0   45   90   135   180 km

Transverse Mercator Projection

Shetland Islands

Orkney Islands

Cape Wrat

John o'Groats

Outer Hebrides
Lewis
Harris
Skye
Mull
Islay

NORTHWEST HIGHLANDS

Great Glen
Loch Ness
River Spey
CAIRNGORMS
River Dee
Aberdeen

1344m Ben Nevis

GRAMPIAN MOUNTAINS
R. Tay
Dundee

SCOTLAND

Loch Lomond
Stirling

Glasgow
R. Clyde
Edinburgh
Firth of Forth

Firth of Clyde
SOUTHERN UPLANDS
R. Tweed

CHEVIOT HILLS

UNITED KINGDOM

NORTHERN IRELAND

ANTRIM MOUNTAINS
R. Bann
Lough Neagh
Belfast

River Erne

North Channel

REPUBLIC OF IRELAND
Lough Corrib
River Shannon

Isle of Man

Newcastle upon Tyne
River Tyne
Sunderland
Stockton-on-Tees
Middlesbrough
NORTH YORK MOORS

LAKE DISTRICT
River Eden
River Tees

PENNINES

River Ouse
York
Kingston-upon-Hull
River Humber

Irish Sea

Blackpool
Bradford
Preston
Leeds
Huddersfield
Bolton
Manchester
Doncaster
Liverpool
Stockport
Warrington
River Mersey
Sheffield
ENGLAND

NORTH ATLANTIC OCEAN

R. Boyne
R. Liffey
Dublin

WICKLOW MOUNTAINS
Barrow
River Suir
River Blackwater
Cork

Anglesey
R. Dee
Stoke-on-Trent
Derby
Nottingham
River Trent
The Wash
R. Wensum
Norwich

Cardigan Bay
CAMBRIAN MOUNTAINS
Telford
Wolverhampton
Walsall
Birmingham
Dudley
Coventry
Solihull
Leicester
Northampton
THE FENS
Peterborough
R. Great Ouse

WALES
River Teifi
River Tywi
River Usk
BRECON BEACONS
Newport
Swansea
Cardiff
R. Wye
R. Severn
R. Avon
COTSWOLD HILLS
Milton Keynes
Luton
R. Stour
Ipswich
CHILTERN HILLS
Basildon
Southend-on-Sea
London
Reading
R. Thames
Bristol
SALISBURY PLAIN
NORTH DOWNS

Bristol Channel

St George's Channel

NORTH ATLANTIC OCEAN

EXMOOR
R. Exe
Southampton
Bournemouth
Poole
Portsmouth
SOUTH DOWNS
Brighton
Strait of Dover
Isle of Wight

DARTMOOR
Plymouth
Torbay

Isles of Scilly
Land's End

English Channel

North Sea

Map of the world

international boundary
• capital city

abbreviations
BELG.	BELGIUM
B-H.	BOSNIA-HERZEGOVINA
C.	CROATIA
CENT. AF. REP.	CENTRAL AFRICAN REPUBLIC
CZ.	CZECH REPUBLIC
F.	FYROM (Former Yugoslav Republic of Macedonia)
K.	KOSOVO
LITH.	LITHUANIA
MT.	MONTENEGRO
LUX.	LUXEMBOURG
NETH.	NETHERLANDS
S.	SLOVENIA
SE.	SERBIA
SL.	SLOVAKIA
SWITZ.	SWITZERLAND
U.A.E.	UNITED ARAB EMIRATES
U.S.A.	UNITED STATES OF AMERICA

Equatorial Scale 1: 95 000 000

The continents and oceans

Amazing – but true!

- Nearly 70% of Earth is covered by saltwater.
- Nearly 1/3 is covered by the Pacific Ocean.
- 10% of the land is covered by glaciers.
- 20% of the land is covered by deserts.

World champions

- Largest continent – Asia
- Longest river – The Nile, Africa
- Highest mountain on land – Everest, Nepal
- Highest mountain in the ocean – Mauna, Hawai
- Largest desert – Sahara, North Africa
- Largest ocean – Pacific

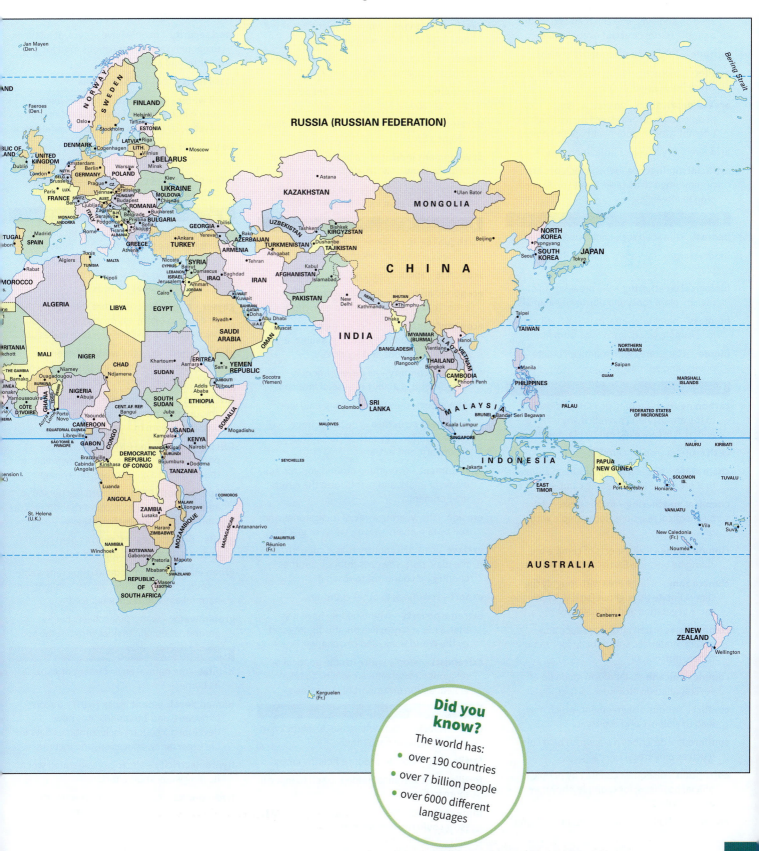

Did you know?

The world has:
- over 190 countries
- over 7 billion people
- over 6000 different languages

Glossary

A

absolute monarch – a monarch who has complete control over his or her people

add value – process something in a way that increases its value; so you can charge more

adult literacy rate – the % of the population aged 15 and over who can read and write a simple sentence about everyday life

altitude – height of a place above sea level

aquifer – a natural underground rock structure which holds groundwater

area of deprivation – an area where people suffer disdvantages such as high unemployment, poor health, poor housing

arid – very dry; receives little or no rain

artificial intelligence – when machines are programmed to do tasks that usually need human intelligence

B

bedrock – the solid rock underlying the soil

biodiversity – the variety of living things in a place

biological weathering – the breaking up of rock by plant roots and burrowing animals

biomass – plant or animal material used as fuel; for example wood, and waste from farms

biome – a large area with a similar climate, plants, and animals

C

carbon neutral – adding no carbon dioxide to the atmosphere, overall; if any is being added, it is cancelled out by planting trees

cash crop – a crop you grow for sale

chain store – one of a group of similar shops that are owned by the same company

chemical weathering – rock is broken down by chemical reactions, for example with rainwater

climate change – all aspects of climate are changing because Earth is getting warmer

climate zone – a large area with roughly the same climate throughout (so it will have the same vegetation and animals too)

commodities – agricultural produce and natural materials, sold in bulk; for example coffee, iron ore, oil

complex volcano – a landform made up of several related volcanoes

conflict – serious disagreement, which may lead to violence and even war

coniferous – describes trees which bear cones (such as pine trees)

corruption – dishonest conduct by people in official positions; for example, they may accept bribes

crater – the hollow around a vent in a volcano

D

deciduous – describes trees which lose their leaves in winter; for example oak trees

decline – to fall gradually into a poor state

deposit – to drop material; waves deposit sand and small stones to form beaches

desalination plant – where seawater is turned into fresh water, by removing its salt

desert – gets very little rain; it can be a hot or cold desert, and sandy or rocky

developing country – its people are poor, on average, and lack many services

development – a process of change to improve people's lives

development indicator – a piece of data that helps to show how developed a country is

diversify – develop a wider range; for example, more ways to earn income

E

earthquake – the shaking of Earth's crust, caused by sudden rock movement

economic – about money, jobs, and business

economic migrant – a person who moves in order to find paid work, or to earn more

economy – all the activities going on in a country, in producing, buying, selling, and distributing goods and services

emerging economy – its development is speeding up, usually because of industry

employment – the state of having paid work

employment structure – the percentage of the workforce in each sector: primary, secondary, and tertiary

environment – everything around you; air, soil, water, animals, and plants form the natural environment

erosion – the wearing away of rock, stones and soil by rain, rivers, waves, wind, or glaciers

exclave – part of a country that is cut off from the main part; you pass through another country (or countries) to reach it

export – sell things to other countries

extreme poverty – where people have less than $1.90 a day to live on; this figure is set by the World Bank, and can change

extremists – people with extreme political or religious views; some are willing to take violent action in support of their views

F

fault – a crack in rock, where blocks of rock can move relative to each other

fertile (soil) – able to produce healthy crops

fertilisers – substances added to soil to make it more fertile

finite resource – there is a limited amount of it (so it could run out one day)

fold mountain – formed by plates pushing into each other; the rock at the plate edges gets folded upwards, making mountains

food bank – a place which gives out free (donated) food to people who need it

food insecurity – people do not have access at all times to the food they need (see next)

food security – people have access, at all times, to enough safe and nutritious food for a healthy and active life

fossil fuels – coal, oil, natural gas

fossil water – groundwater that has been sealed into aquifers for thousands or millions of years, by changes in the surrounding rock

fresh water – water that is not salty; for example the water in rivers and springs

fumarole – a vent or opening in Earth's crust which emits steam and gases

G

geopolitics – how human and geographical factors shape the relationships between countries; for example a country may want to stay friendly with a country that borders it, or has resources it needs

geyser – a spring of water that shoots into the air every so often; it is heated by hot rock below ground, and the pressure builds up

globalisation – how the world is becoming more interconnected, through movement of goods, people, money, and information

GNI (gross national income) – the total amount that a country's population and businesses earn in a year

GNI per person – the GNI divided by the population; it is a measure of how wealthy the people in a country are, on average

GNI per person (PPP) – the GNI per person is adjusted to take into account that things cost more in some places than others

goods – physical objects, such as pens, apples, and cars, that are bought and sold

green electricity – generated using renewable resources such as wind, waves, sunlight; no harmful emissions are produced

groundwater – water that collects below ground, when rain trickles through the soil

H

habitat – the natural environment of a species; its home

human development index (HDI) – a score between 0 and 1 that indicates how developed a country is; it combines data on life expectancy, education, and income

hydroelectricity – electricity generated when flowing water spins a turbine

hydroponics – growing plants in water that contains the nutrients they need

I

igneous rock – forms when melted rock hardens

import – buy in things from other countries

Industrial Revolution – the period (about 1760 – 1840) when many new machines were invented in the UK, and factories built

inequality – when wealth and access to services are not shared equally

infrastructure – facilities such as roads, water supply, electricity grid, and railways, that keep a country and its economy running

irrigate – to water crops

L

landlocked – surrounded by land, with no coastline

latitude – how far a place is north or south of the Equator; it is measured in degrees

latrine – a very basic toilet; it could be just a hole, or trench, in the ground

lava – melted rock at Earth's surface

life expectancy – how many years a new baby can expect to live for, on average

living sustainably – living in a way that does not harm us humans, other species, or the environment

longitude – how far a place is east or west of the Prime Meridian; it is measured in degrees

M

magma – melted rock below Earth's surface

magnitude – how much energy an earthquake gives out (measured on the Richter scale)

manufacturing – making things in factories

mass extinction – when a large number of species die off over quite a short period of time; for example because of an ice age

metamorphic rock – forms when rock is changed through the action of heat and / or pressure, without melting

migrant – a person who moves somewhere else, usually to find work, or a better life

mineral – a natural compound in rock

N

natural resource – it occurs naturally in the environment, and we can make use of it; for example, wind and oil

NGO (non-governmental organisation) – a charity that helps people, and is not linked to a government; for example, Oxfam

non-renewable resource – a resource that is limited, and could run out one day

North Atlantic Drift – a warm current in the Atlantic Ocean; it flows up to the Arctic

nutrients (for plants) – substances which plants need, to grow and be healthy

O

out-of-town shopping centre – a centre built for shopping, outside a town or city, with a range of shops, and free car parking

P

pandemic – a disease that spreads over a very wide area, or the whole world, and affects a large number of people

peninsula – land that juts out into the sea, and is almost surrounded by water

permafrost – the ground under the surface that is permanently frozen, in the tundra

permit – a document giving a person permission to do something

physical weathering – weathering which breaks rock into smaller bits

plateau – an area of fairly flat high land

plates – Earth's hard outer part is broken into big slabs called plates, which move around

populous – has a large population

poverty – the state of being poor

poverty line – where a sum of money is used to define the level of poverty in a country; for example, the % of people living on less than $1.90, or $5.50, a day

precipitation – water falling from the sky (as rain, sleet, hail, snow)

primary products – goods like oil and timber, from the primary sector of the economy

primary sector – the sector of the economy where people collect things from Earth; farmers and miners are in this sector

PV cell (photovoltaic cell) – converts the energy from sunlight into electricity

Q

quality of life – the level of comfort and well-being a person enjoys

quaternary sector – the sector of the economy where people use high-level expertise to develop things that will help other sectors

R

refugee – a person who has been forced to flee from danger; for example from war

relief – how the height of the land varies

renewable resource – a resource we can make use of, that will not run out; for example, sunlight and wind

replacement fertility rate – the average number of children per woman that keeps a population at the same size (without migration); it is taken as 2.1 children

republic – does not have a monarch

resources – things we need to live, or use to earn a living; for example food, fuel

Ring of Fire – the ring of volcanoes and earthquake sites circling the Pacific Ocean

S

sanctions – penalties a country places on another country, to punish it; for example it may refuse to buy goods from that country

sedimentary rock – formed from sediment; sandstone is formed from a sediment of sand

secondary sector – the sector of the economy where people build things and make things, for example in factories

services – activities carried out to meet needs, such as teaching, or treating tooth decay

social – about people and society

solar power – electricity generated from sunlight, using photovoltaic (PV) cells

standard of living – the level of goods, services, and comfort available to people

steppe – a large flat area of treeless grassland

sustainable – can be carried on into the future without doing harm

T

taiga – region of coniferous forests which lies between the tundra and steppe

tariffs – taxes on imports or exports

temperate – term used for a mild climate

tertiary sector – the sector of the economy where people provide services for other people; for example healthcare

trade war – where countries try to damage each other's trade; for example, they may place tariffs on imports from each other

transnational corporation (TNC) – a company that operates in more than one country; examples are Apple, Nike, Nissan

tree line – the line or altitude above which it is too cold for trees to grow

tsunami – waves generated by an earthquake in the ocean floor

tundra – a cold region where the ground is deeply frozen; only the surface thaws in summer, allowing small plants to grow

U

under-5 mortality rate – the % of babies born alive who die before age 5

unemployment – when people are looking for paid work, and can't find it

utilities – services provided, such as electricity, gas, and water supplies, sewage removal, and phone lines

V

volcano – where lava erupts at Earth's surface

W

water stress – where a country or area cannot meet its demand for fresh water

weathering – the breaking down of rock, by the action of things in its environment: heat, cold, rain, plant roots, and so on

Index